CHEMISTRY CALCULATIONS

J. A. Hunt M.A.
Durrants School, Croxley Green

A. Sykes B.Tech., Ph.D.
Formerly of Alleyne's School, Stevenage

Longman

Longman Group UK Limited
*Longman House, Burnt Mill, Harlow, Essex CM20 2JE, England
and Associated Companies throughout the World*

First published 1985
Eighth impression 1994
ISBN 0 582 33181 1

Set in Monophoto Times New Roman
The publisher's policy is to use paper manufactured from sustainable forests.

Produced by Longman Singapore Publishers Pte Ltd
Printed in Singapore

Acknowledgements

We are grateful to the following Examining Bodies for permission to reproduce questions from past examination papers:

The Associated Examining Board (**AEB**); Associated Lancashire Schools Examining Board (**ALSEB**); East Anglian Examinations Board (**EAEB**); East Midland Regional Examinations Board (**EMREB**); Joint Matriculation Board (**JMB**); North West Regional Examinations Board (**NWREB**); Northern Ireland Schools Examinations Council (**NISEC**); Oxford & Cambridge Schools Examination Board (**O & C**); Southern Universities Joint Board (**SUJB**); University of Cambridge Local Examinations Syndicate (**CLES**); University of London School Examinations Department (**L**); University of Oxford Delegacy of Local Examinations (**O**); Welsh Joint Education Committee (**WJEC**).

We would like to take this opportunity to accept full responsibility for the answers given to the examination questions in this book. These answers were neither provided nor approved by the examination boards concerned.

Contents

Preface

This book is designed to complement our textbook *Chemistry*. We have written it primarily to help those preparing for public examinations (CSE, GCE and GCSE) at the age of sixteen. The questions at the end of each chapter illustrate the variety and importance of numerical and graphical problems in chemistry examinations at this level. We hope it will also help students starting A level courses in chemistry, especially those who have previously followed science syllabuses with little emphasis on chemical calculations.

We have used the 'mole' as the unifying concept for this book. This has the advantage of reducing the number of separate ideas needed to solve quantitative problems. It ceases to be necessary to mention Faraday's laws or Gay Lussac's law explicitly, despite their great importance in the history of chemistry.

The supporting data needed for calculations throughout the book is listed inside the back cover.

We have used the same conventions for units as in *Chemistry*. The examination boards, however, use a variety of conventions, so there are some inconsistencies between the main text and the examination questions at the ends of the chapters. The worked examples show how calculations should be set out step by step, with the units included at each stage. The use of calculators may produce numerical answers which are much more precise than is justified by the data. The final answer to a question should be rounded to the number of significant figures consistent with the information given.

We should like to take this opportunity to thank all those who helped us with the preparation of this book. The cartoons were devised and drawn by Sonja Rivers while she was a sixth-form student at Durrants School. Julia Huyton at Alleyne's School gave us the idea of developing sets of rules to clarify the steps of calculations. John Griffiths at Watford Grammar School offered valuable advice about the style of the mathematics in the worked examples. Barry Nelson at North Westminster Community School read the typescript and made helpful suggestions to improve its chemical content. Finally we should like to thank all those at Longman who have worked to prepare this book for publication and especially Laurice Suess for her invaluable contribution to the clarity and accuracy of the text.

J. A. Hunt
A. Sykes
January 1985

1 Moles

1.1 Atomic structure

The mass of an atom is concentrated in its nucleus, which is very small. The nucleus is made of protons and neutrons. The protons are positively charged. The neutrons are uncharged. Protons and neutrons have the same mass. Around the nucleus are the electrons, which are negatively charged. The mass of the electrons is so small that it can be ignored in most calculations. The diagram in figure 1.1 shows the number and arrangement of the particles in atoms of hydrogen, helium and lithium.

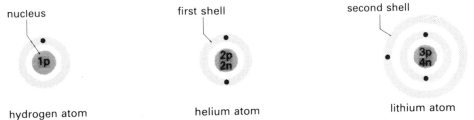

nucleus   first shell   second shell

hydrogen atom   helium atom   lithium atom

Figure 1.1. Models of hydrogen, helium and lithium atoms: protons, neutrons and electrons are shown by p, n and ·

Atoms differ in mass because of the differing numbers of protons and neutrons. The total number of protons and neutrons in the nucleus is called the *mass number*, A, of the atom. The number of protons is called the *atomic number*, Z. The number of electrons in an atom is equal to the number of protons. Atoms are uncharged because the total negative charge on the electrons balances the total positive charge on the protons.

A shorthand for describing atoms is as follows:

mass number—$_A$
atomic number—$_Z$ X — symbol

The symbols for the atoms in figure 1.1 are: ^1_1H, ^4_2He and ^7_3Li.

All atoms of the same element have the same atomic number, but the number of neutrons can vary. There are three types of silicon atom: $^{28}_{14}Si$, $^{29}_{14}Si$ and $^{30}_{14}Si$. These are the three *isotopes* of silicon. They are all silicon atoms because they contain fourteen protons.

Atoms can become electrically charged by gaining or losing electrons. Charged atoms are called *ions*. Non-metal atoms, such as chlorine atoms, gain electrons and form negative ions. Metal atoms, such as sodium atoms, lose electrons and become positive ions.

Example 1a

How many protons, neutrons and electrons are there in the atom $^{19}_9F$?

Answer

The number of protons = Atomic number
　　　　　　　　　　　 = 9

The number of neutrons = Mass number − Atomic number
　　　　　　　　　　　　 = 19 − 9
　　　　　　　　　　　　 = 10

The number of electrons = Number of protons
　　　　　　　　　　　　 = 9

Example 1b

How many protons, neutrons and electrons are there in the ion $^{137}_{56}Ba^{2+}$?

Answer

The number of protons = 56

The number of neutrons = 137 − 56
　　　　　　　　　　　　 = 81

The ion is positively charged because it has lost two electrons so that there are not enough electrons to balance the charge on the nucleus

The number of electrons = 56 − 2
　　　　　　　　　　　　 = 54

Questions

1　How many protons, neutrons and electrons are there in these particles: (a) ^9_4Be, (b) $^{39}_{19}K$, (c) $^{235}_{92}U$, (d) $^{127}_{53}I^-$, (e) $^{40}_{20}Ca^{2+}$?

2　Write the symbols, showing the mass number and atomic number, for these atoms or ions:

(a) an atom of oxygen containing 8 protons, 8 neutrons and 8 electrons,

(b) an atom of argon containing 18 protons, 22 neutrons and 18 electrons,

(c) an ion of sodium which has a 1+ charge and a nucleus made up of 11 protons and 12 neutrons,

(d) an ion of sulphur which has a 2− charge and a nucleus made up of 16 protons and 16 neutrons.

1.2 Relative atomic mass

A hydrogen atom is the lightest of all atoms. It consists of one proton and one electron. Electrons have negligible mass, so the mass of a hydrogen atom is almost the same as that of one proton. The *relative atomic mass* of hydrogen is 1.

All other atoms are heavier than hydrogen atoms. Oxygen has a relative atomic mass of 16. Each oxygen atom is sixteen times heavier than a hydrogen atom.

The relative atomic mass of sulphur is 32. So a sulphur atom is thirty-two times heavier than a hydrogen atom but twice as heavy as an oxygen atom.

A_r is the accepted symbol for relative atomic mass. $A_r(H) = 1$, $A_r(O) = 16$ and $A_r(S) = 32$.

Questions

Use the table of relative atomic masses inside the back cover of this book to answer these questions.

3 (a) These elements are listed in alphabetical order: aluminium, argon, chlorine, magnesium, phosphorus, silicon, sodium, sulphur. Arrange them in order of relative atomic mass, starting with the lightest.

(b) What is the connection between your answer to part (a) and the periodic table?

4 How many times heavier than hydrogen atoms are the atoms of

(a) carbon, (b) magnesium, (c) bromine, (d) barium, (e) lead?

5 How many times heavier is

(a) a magnesium atom than a carbon atom?

(b) a nitrogen atom than a lithium atom?

(c) a sulphur atom than a helium atom?

(d) a bromine atom than an argon atom?

(e) an iron atom than a nitrogen atom?

Relative atomic masses are not always whole numbers. The relative atomic mass of chlorine is 35.5. This is because chlorine is a mixture of two isotopes: chlorine-35 and chlorine-37. On average, out of every four atoms, three are chlorine-35 and one is chlorine-37.

The average relative atomic mass of chlorine $= \dfrac{(3 \times 35) + (1 \times 37)}{4}$

$$= 35.5$$

Questions

6 In 100 atoms of gallium, 60 atoms are gallium-69 and 40 atoms are gallium-71. What is the relative atomic mass of gallium?

7 Antimony consists of 57% antimony-121 and 43% antimony-123. What is the relative atomic mass of antimony?

1.3   Moles of atoms

12 g of carbon contain the same number of atoms as 1 g of hydrogen because each carbon atom is twelve times heavier than each hydrogen atom. Similarly, 16 g of oxygen ($A_r = 16$) and 32 g of sulphur ($A_r = 32$) contain the same number of atoms as 12 g of carbon or 1 g of hydrogen. The amount of an element which contains the same number of atoms as 1 g of hydrogen or 12 g of carbon is called *one mole* of that element.

The mass in grams of one mole of an element is numerically equal to its relative atomic mass. For example, the mass of one mole of oxygen ($A_r = 16$) is 16 g.

Every unit has a name and a symbol. A volume of six cubic centimetres is written 6 cm³. A mass of six grams is written 6 g. Similarly, an amount of six moles is written 6 mol.

Example 1c
What is the mass of 5 mol of fluorine atoms?

Answer
The table inside the back cover shows that $A_r(F) = 19$
The mass of 1 mol of fluorine atoms = 19 g

The mass of 5 mol of fluorine atoms = 5 × 19 g
$$= 95\,g$$

Example 1d
How many moles of atoms are there in 1.6 g of copper?

Answer
The table inside the back cover shows that $A_r(Cu) = 64$
The mass of 1 mol of copper = 64 g
1.6 g of copper must be a small fraction of a mole

The number of moles of copper in 1.6 g of the metal $= \dfrac{1.6\,g}{64\,g}$
$$= 0.025$$

Questions

8 What is the mass of
   (a) 1 mol of sodium atoms?
   (b) 10 mol of chlorine atoms?
   (c) 0.1 mol of iodine atoms?
   (d) 0.5 mol of iron atoms?
   (e) 0.125 mol of bromine atoms?

9 How many moles of atoms are there in
   (a) 27 g of aluminium?
   (b) 20 g of calcium?
   (c) 4 g of bromine?
   (d) 140 g of nitrogen?
   (e) 0.65 g of zinc?

10 (a) What mass of carbon contains the same number of atoms as 39 g of potassium?
   (b) What mass of sulphur contains the same number of atoms as 6 g of magnesium?
   (c) What mass of copper contains the same number of atoms as 2.07 g of lead?
   (d) What mass of fluorine contains the same number of atoms as 4000 g of helium?

11 (a) How many moles of lithium contain the same number of atoms as 320 g of oxygen?
   (b) How many moles of phosphorus contain the same number of atoms as 2.0 g of neon?
   (c) How many moles of tin contain the same number of atoms as 0.201 g of mercury?

1.4 Relative molecular mass and moles of molecules

Most non-metal elements consist of molecules. For example: oxygen, O_2; sulphur, S_8; hydrogen, H_2; and phosphorus, P_4.

Most compounds of non-metals with non-metals are also molecular. Examples include water (H_2O), carbon dioxide (CO_2), hydrogen chloride (HCl) and silicon tetrachloride ($SiCl_4$).

The *relative molecular mass* of a non-metal element, or of a compound of non-metals, is found by adding up the relative atomic masses in the molecular formula.

Example 1e
What is the relative molecular mass of ethanol, C_2H_6O?

Answer

The relative molecular mass of ethanol $= (2 \times 12) + (6 \times 1) + 16$

$$A_r(C) \qquad A_r(H) \quad A_r(O)$$

$$= 46$$

The example shows that a molecule of ethanol is 46 times heavier than one atom of hydrogen. This means that in 46 g of ethanol there is the same number of molecules as there are atoms in 1 g of hydrogen. The mass of one mole of ethanol molecules is 46 g.

The mass of one mole of molecules of a substance is numerically equal to its relative molecular mass, if the mass is measured in grams.

Example 1f

What is the mass of 2 mol of sulphuric acid, H_2SO_4?

Answer

The relative molecular mass of sulphuric acid $= (2 \times 1) + 32 + (4 \times 16)$
$$= 98$$

The mass of 1 mol of sulphuric acid $\qquad = 98$ g

The mass of 2 mol of sulphuric acid $\qquad = 2 \times 98$ g
$$= 196 \text{ g}$$

Questions

12 What is the relative molecular mass of
  (a) oxygen, O_2?
  (b) carbon dioxide, CO_2?
  (c) hydrogen iodide, HI?
  (d) silicon tetrachloride, $SiCl_4$?

13 What is the mass of
  (a) 1 mol of chlorine molecules, Cl_2?
  (b) 2 mol of water molecules, H_2O?
  (c) 0.5 mol of sulphur molecules, S_8?
  (d) 0.25 mol of phosphorus molecules, P_4?

14 How many moles of molecules are there in
  (a) 6.4 g of sulphur dioxide, SO_2?
  (b) 56 g of carbon monoxide, CO?
  (c) 14 g of nitrogen, N_2?
  (d) 160 g of bromine, Br_2?
  (e) 4.0 g of hydrogen fluoride, HF?

15 What mass of
  (a) oxygen contains the same number of molecules as 8.0 g of sulphur?
  (b) water contains the same number of molecules as 22 g of carbon dioxide?

1.5 Relative formula mass and moles of salts

Salts are ionic compounds. They are usually compounds of metals with one or more non-metals. They are made up of a giant structure of ions. Examples are calcium chloride, $CaCl_2$, lead sulphate, $PbSO_4$, and magnesium nitrate, $Mg(NO_3)_2$.

The *relative formula mass* of a salt is found by adding up the relative atomic masses in the formula. Some salts include water of crystallization and this must be counted in the relative formula mass if it is shown in the formula.

Example 1g
What is the relative formula mass of magnesium nitrate, $Mg(NO_3)_2$?

Answer
The relative formula mass $= 24 + 2 \times (14 + 48)$
$$= 148$$

Example 1h
What is the relative formula mass of blue copper(II) sulphate crystals, $CuSO_4 \cdot 5H_2O$?

Answer
The relative formula mass of anhydrous copper(II) sulphate, $CuSO_4$
$$= 64 + 32 + (4 \times 16)$$
$$= 160$$
The relative molecular mass of water
$$= (2 \times 1) + 16$$
$$= 18$$
The relative formula mass of hydrated copper(II) sulphate
$$= 160 + (5 \times 18)$$
$$= 250$$

When doing these calculations, remember the rules for the order of mathematical operations. In the first example above, to work out

$24 + 2 \times (14 + 48)$

you work out the parts in brackets first, do divisions next, then multiplications, then additions and finally subtractions. The word BODMAS reminds you of this order.

Questions
16 What is the relative formula mass of
    (a) potassium chloride, KCl?
    (b) aluminium bromide, $AlBr_3$?

(c) iron(III) sulphate, $Fe_2(SO_4)_3$?
(d) zinc sulphate, $ZnSO_4 \cdot 7H_2O$?
(e) sodium carbonate, $Na_2CO_3 \cdot 10H_2O$?
17 What is the mass of
(a) 1 mol of zinc sulphide, ZnS?
(b) 0.5 mol of lead nitrate, $Pb(NO_3)_2$?
(c) 5 mol of sodium sulphate, Na_2SO_4?
(d) 0.1 mol of iron(II) sulphate, $FeSO_4 \cdot 7H_2O$?
(e) 0.01 mol of ammonium chloride, NH_4Cl?

In calcium chloride, $CaCl_2$, there are two chloride ions, Cl^-, combined with each calcium ion, Ca^{2+}. So in one mole of calcium chloride there are two moles of chloride ions.

At first sight this is a confusing idea. Think about eggs in boxes. Each egg box contains six eggs. In a dozen egg boxes there are six dozen eggs. In one hundred egg boxes there are six hundred eggs. Similarly, in one mole of egg boxes there are six moles of eggs. Compare the formula of a salt to the egg box, and the ions inside to the eggs. In the formula of aluminium chloride, $AlCl_3$, one aluminium ion is combined with three chloride ions. In one mole of aluminium chloride there are three moles of chloride ions.

Example 1i
How many moles of (a) calcium ions and (b) phosphate ions are there in three moles of calcium phosphate, $Ca_3(PO_4)_2$?

Answer
In one mole of $Ca_3(PO_4)_2$ there are three moles of calcium ions, $3Ca^{2+}$, and two moles of phosphate ions, $2PO_4^{3-}$. So in three moles of calcium phosphate there are

$$3 \times 3 = 9 \, \text{mol of calcium ions}$$

and $3 \times 2 = 6 \, \text{mol of phosphate ions}$

Question
18 (a) How many moles of sodium ions are there in 1 mol of sodium carbonate, Na_2CO_3?
(b) How many moles of nitrate ions are there in 2 mol of magnesium nitrate, $Mg(NO_3)_2$?
(c) How many moles of bromide ions are there in 0.5 mol of barium bromide, $BaBr_2$?
(d) How many moles of nitrogen atoms are there in 1 mol of ammonium sulphate, $(NH_4)_2SO_4$?
(e) How many moles of water molecules are there in 5 mol of blue copper(II) sulphate, $CuSO_4 \cdot 5H_2O$?

1.6   Moles into masses and masses into moles

This equation is useful when answering questions of the type set in this chapter:

✽ Mass of a substance (g) = Number of moles × Mass of one mole (g)

Use the equation in this form to answer question **19**.
   The equation can be rearranged as follows:

✽ Number of moles of a substance = $\dfrac{\text{Mass of the substance (g)}}{\text{Mass of one mole (g)}}$

Use the equation in this form to answer question **20**.

Questions
19 What is the mass of
   (a) 1 mol of iron(II) chloride, $FeCl_2$?
   (b) 2 mol of manganese(II) bromide, $MnBr_2$?
   (c) 0.5 mol of nitric acid, HNO_3?
   (d) 0.002 mol of ethane molecules, C_2H_6?
   (e) 10 mol of gallium atoms?
   (f) 0.25 mol of sulphur molecules, S_8?
   (g) 0.125 mol of sulphate ions, SO_4^{2-}?
20 How many moles of
   (a) atoms are there in 36 g of beryllium?
   (b) atoms are there in 14 g of iron?
   (c) molecules are there in 11 g of carbon dioxide, CO_2?
   (d) molecules are there in 7 g of ethene, C_2H_4?
   (e) sodium ions are there in 40 g of sodium hydroxide, $NaOH$?
   (f) carbonate ions are there in 50 g of calcium carbonate, $CaCO_3$?

1.7   The Avogadro constant

The number of particles in one mole is called the *Avogadro constant*. The
number is approximately 600 000 000 000 000 000 000 000 particles per mole. The
symbol for the Avogadro constant is L. Thus

   $L = 6 \times 10^{23}$ per mole

It is important to state clearly which particles are involved. In 18 g of water
there are 6×10^{23} water molecules, H_2O, made from 12×10^{23} hydrogen atoms
joined to 6×10^{23} oxygen atoms.

Questions
21 Use the Avogadro constant to work out the number of oxygen atoms in
   (a) two moles of oxygen molecules, O_2.
   (b) one mole of sulphur dioxide, SO_2.

(c) four moles of sulphuric acid, H_2SO_4.
(d) one mole of blue copper(II) sulphate crystals, $CuSO_4 \cdot 5H_2O$.

22 Use the Avogadro constant to calculate the number of
(a) chloride ions in one mole of magnesium chloride, $MgCl_2$.
(b) sulphur atoms in two moles of rhombic sulphur, S_8.
(c) sodium ions in one mole of sodium carbonate, Na_2CO_3.
(d) sulphate ions in three moles of aluminium sulphate, $Al_2(SO_4)_3$.

Examination questions

The ideas in this chapter are usually tested by multiple choice questions in public examinations. Questions **1–18** below are short calculations that are similar to the ones that are usually included in multiple choice tests. (There is a revision multiple choice test on pp. 155–62, which covers all the ideas in this book.) Questions **19–24** are taken from papers of the examination boards indicated.

1   How many protons and neutrons are there in the nucleus of a sodium atom, $^{23}_{11}Na$?
2   How many electrons are there in a helium atom, ^4_2He?
3   The atomic number of chromium is 24 and its mass number is 52. How many neutrons are there in a chromium atom?
4   What mass of magnesium contains the same number of atoms as 8.0 g of calcium?
5   How many moles of magnesium atoms are there in 0.2 g of the metal?
6   What mass of magnesium contains the same number of atoms as 7.0 g of iron?
7   What mass of sulphur, S_8, contains the same number of molecules as there are atoms in 12.0 g of carbon?
8   How many moles of oxygen atoms are there in 6.3 g of nitric acid, HNO_3?
9   What is the mass of two moles of anhydrous sodium sulphate, Na_2SO_4?
10   What is the mass of 0.004 mol of sodium hydrogensulphate, $NaHSO_4$?
11   What is the mass of 0.002 mol of magnesium carbonate, $MgCO_3$?
12   What is the number of moles of hydrogen atoms in 6.4 g of methane, CH_4?
13   How many moles of chlorine gas, Cl_2, contain the same number of atoms as there are in 16 g of helium?
14   How many moles of chloride ions are there in a mixture of 2 mol of barium chloride, $BaCl_2$, and 1 mol of potassium chloride, KCl?
15   On heating 0.56 g of copper(II) bromide, it decomposed to give 0.36 g of copper(I) bromide and some bromine. How many moles of bromine molecules, Br_2, were given off?
16   4.68 g of a crystalline solid was heated and water vapour was evolved. The mass of the residue was 3.96 g. How many moles of water molecules, H_2O, were given off during heating?
17   If the number of particles in one mole is 6×10^{23}, how many oxygen molecules, O_2, are there in 1.6 g of oxygen gas?
18   Use the Avogadro constant to calculate the number of sodium ions in 6.2 g of sodium oxide, Na_2O.

19 How many moles of atoms are there in 4 g of each of the following elements:
(a) carbon,   (b) magnesium,   (c) sulphur,   (d) calcium?   **(SUJB)**

20 (a) Complete the following table.

	Mass number	Atomic number of element	Number of		
			electrons	protons	neutrons
Sodium atom	23	11			
Calcium ion	40		18		
Chloride ion		17			20

(b) What name is used to describe atoms which have the same atomic number but different mass numbers?

(c) How many electrons, protons and neutrons are present in the hydrogen ion, H^+?   **(C)**

21 Chlorine consists of two isotopes, $^{35}_{17}Cl$ and $^{37}_{17}Cl$.
(a) How do the nuclei of these isotopes differ?
(b) Explain why the relative atomic mass of natural chlorine is 35.5.
(c) Give the electronic structure of atoms of each isotope.
(d) Give the three relative masses you would expect for molecules of natural chlorine, stating which you think would be the most common.   **(O)**

22 The symbols $^{20}_{10}Ne$ and $^{22}_{10}Ne$ represent *isotopes* of neon.
(a) Explain the meaning of *isotopes*.
(b) Give the composition of the nuclei of the two atoms of neon.
(c) How do you account for the fact that the relative atomic mass of naturally occurring neon is 20.2?   **(WJEC)**

23 'Yellow metal' is an alloy containing 60% copper and 40% zinc by weight. Calculate the number of moles of zinc atoms in 260 g of the alloy.   **(NISEC)**

24 12 g of carbon contains 6×10^{23} atoms of carbon.
(a) How many atoms would there be in 1 g of carbon?
(b) How many atoms of hydrogen would there be in 1 g of hydrogen?   **(EAEB)**

2 Moles and formulae

2.1 Finding formulae

Chemical formulae are first found by experiment. This has been done for all the common compounds and you can look up their formulae in tables of data. However, you are expected to understand how formulae can be worked out by experiment and calculation.

An experiment to find a formula involves measuring the masses of the elements which combine in the compound. From this information the *empirical formula* is calculated. This shows the *ratio* between the numbers of each type of atom in the compound. If the substance is molecular, further experiments are necessary to determine the *molecular formula*, which shows how *many* atoms of each type there are in a molecule.

Empirical formulae can be found from experimental results either by using a graph or by calculation.

The graphical method
This method only works for compounds of just two elements. It is a good way of averaging a set of experimental results. To use this method, it is necessary to guess what the formula is likely to be before testing the guesses graphically.

Example 2a

A class carried out a series of experiments to find the formula of lead bromide. The results are shown in figure 2.1. What is the formula of lead bromide?

Mass of lead (g)	0.16	0.38	0.64	0.92	1.56	1.80
Mass of bromine (g)	0.13	0.34	0.54	0.74	1.14	1.44

Figure 2.1

Answer

The results are first plotted on a graph, as shown in figure 2.2.

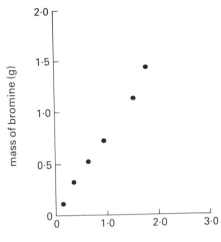

Figure 2.2   mass of lead (g)

All the experimental points lie close to a straight line. This is because lead bromide has a definite formula. The mass of bromine increases in proportion to the mass of lead. Double the mass of lead, and you double the mass of bromine which combines with it.

Possible formulae for lead bromide might be $PbBr$, $PbBr_2$ and Pb_2Br. These can be tested as follows.

The relative atomic masses are $A_r(Pb) = 207$ and $A_r(Br) = 80$.

If the formula is $PbBr$:

        1 mol of lead atoms combines with 1 mol of bromine

hence   207 g of lead combines with 80 g of bromine

This can be scaled down to the amounts used in the experiment by dividing by 100:

        2.07 g of lead combines with 0.80 g of bromine

If the formula is $PbBr_2$:

        1 mol of lead atoms combines with 2 mol of bromine

hence   207 g of lead combines with 160 g of bromine

which on scaling becomes

        2.07 g of lead combining with 1.60 g of bromine

If the formula is Pb_2Br:

        2 mol of lead atoms combines with 1 mol of bromine

hence   414 g of lead combines with 80 g of bromine

which on scaling becomes

        4.14 g of lead combining with 0.80 g of bromine

or   2.07 g of lead combining with 0.40 g of bromine

If the formula is PbBr, all the experimental points will lie on, or near, a straight line from (0,0) through (2.07, 0.80).

If the formula is $PbBr_2$, all the points will be on the line from (0,0) through (2.07, 1.60).

If the formula is Pb_2Br, the points will be on the line from (0,0) through (2.07, 0.40).

These three possible lines are drawn in on the graph in figure 2.3.

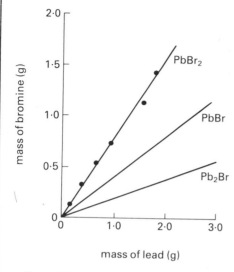

Figure 2.3

Clearly, the formula of lead bromide is $PbBr_2$.

Question

1   The results in figure 2.4 were obtained in a series of experiments to find the formula of sodium oxide. Use the graphical method to find the formula of sodium oxide. Test these possibilities: Na_2O, NaO, NaO_2.

Mass of sodium (g)	0.10	0.15	0.22	0.29	0.33	0.41
Mass of oxygen (g)	0.04	0.05	0.08	0.10	0.12	0.14

Figure 2.4

Calculation method

It is usually easier and quicker to work out formulae by calculation. In this method there is no need to guess the answer beforehand. The rules for finding formulae are shown in figure 2.5.

RULES

FINDING FORMULAE

1/ Write down the masses of elements combining.
2/ Write down the mass of one mole of each element.
3/ Work out amounts (numbers of moles) combining.
4/ Work out the simplest ratio between the amounts combining.
5/ Write down the simplest formula.

Figure 2.5

Example 2b
What is the formula of sodium oxide if 0.69 g of sodium combines with 0.24 g of oxygen?

Answer

	Sodium	**Oxygen**	
Masses combining	0.69 g	0.24 g	(from the information in the question)
Mass of one mole of the element	23 g	16 g	(from tables of data)
Number of moles combining	$\dfrac{0.69\,g}{23\,g} = 0.030$	$\dfrac{0.24\,g}{16\,g} = 0.015$	(divide the mass of each element by the mass of one mole)
Simplest ratio of numbers of moles	2	1	(divide by the smaller number to get the ratio)

The formula is Na_2O

Question

2 Work out the formula of
 (a) magnesium nitride, in which 3.6 g of magnesium combines with 1.4 g of nitrogen.
 (b) methane, given that in 0.8 g of the gas there is 0.6 g of carbon and the rest is hydrogen.
 (c) silicon oxide, given that 6.0 g of the oxide contains 2.8 g of silicon.
 (d) iron bromide, if 0.378 g of iron reacts with bromine to form 2.00 g of the compound.

This method also works for compounds of more than two elements. All that is necessary is to add extra columns to the table. In step 4, the simplest ratio is found by dividing through by the smallest number of moles that was found in step 3.

Example 2c

What is the formula of a compound in which 1.30 g of zinc combines with 0.24 g of carbon and 0.96 g of oxygen?

Answer

	Zinc	**Carbon**	**Oxygen**
Masses combining	1.30 g	0.24 g	0.96 g
Mass of one mole of each element	65 g	12 g	16 g
Number of moles combining	$\dfrac{1.30\,g}{65\,g} = 0.02$	$\dfrac{0.24\,g}{12\,g} = 0.02$	$\dfrac{0.96\,g}{16\,g} = 0.06$
Simplest ratio of the numbers of moles	1	1	3

The formula is $ZnCO_3$

Question

3 What is the formula of a compound in which
   (a) 10 g of calcium combines with 3 g of carbon and 12 g of oxygen?
   (b) 1.38 g of sodium combines with 0.96 g of sulphur and 1.92 g of oxygen?
   (c) 0.48 g of carbon combines with 0.08 g of hydrogen and 0.64 g of oxygen?
   (d) 0.98 g of nitrogen combines with 0.21 g of hydrogen and 1.12 g of oxygen?

The results of analysing a compound are often given in the form of its *percentage composition by mass.* For example, the analysis of copper pyrites shows that it consists of 34.6% copper, 30.5% iron and 34.9% sulphur. This means that in 100 g of the compound there are 34.6 g of copper, 30.5 g of iron and 34.9 g of sulphur. So the formula can be worked out as in the previous examples.

Example 2d

What is the formula of copper pyrites, which has the analysis, 34.6% Cu, 30.5% Fe, 34.9% S?

Answer

	Copper	**Iron**	**Sulphur**
Masses combining	34.6 g	30.5 g	34.9 g
Mass of one mole	64 g	56 g	32 g

Number of moles combining	$\dfrac{34.6\,\text{g}}{64\,\text{g}} = 0.54$	$\dfrac{30.5\,\text{g}}{56\,\text{g}} = 0.54$	$\dfrac{34.9\,\text{g}}{32\,\text{g}} = 1.09$
Simplest ratio	1	1	2

The formula is $CuFeS_2$

Question

4 What is the empirical formula of
   (a) a carbohydrate containing 40.00% carbon, 6.66% hydrogen and 53.33% oxygen?
   (b) a hydrocarbon that contains 82.73% carbon?
   (c) an alcohol with the analysis, 52.2% C, 13.0% H, 34.8% O?
   (d) a mineral acid with the analysis, 2.04% H, 32.46% S, 65.4% O?

2.2 Types of formula

The formula found by experiment is called the *empirical formula*. It shows the simplest ratio between the numbers of each type of atom in a molecule or giant structure. The empirical formula of ethanoic acid is CH_2O.

The *molecular formula* shows how many atoms of each type there are in a molecule. The molecular formula is a simple multiple of the empirical formula. The molecular formula can be found by measuring the relative molecular mass of the compound. The relative molecular mass of ethanoic acid is 60, which corresponds to $(CH_2O)_2$. So the molecular formula of ethanoic acid is $C_2H_4O_2$.

The *structural formula* shows the bonds between the atoms in a molecule. The structural formula of ethanoic acid is:

$$
\begin{array}{ccc}
\text{H} & & \text{O} \\
| & & \parallel \\
\text{H}-\text{C}-\text{C} & & \\
| & & \diagdown \\
\text{H} & & \text{O}-\text{H}
\end{array}
$$

Possible structures can be worked out from the numbers of covalent bonds that atoms normally form in molecules. These are shown in figure 2.6.

Element	Symbol	Usual number of covalent bonds formed in molecules
Carbon	C	4
Hydrogen	H	1
Oxygen	O	2
Nitrogen	N	3
Chlorine	Cl	1

Figure 2.6

Example 2e

Analysis of a hydrocarbon shows that it consists of 81.8 % carbon. The relative molecular mass of the compound is 44. Calculate the empirical formula of the compound. What is the molecular formula? Draw a possible structure for the compound.

Answer

	Carbon	**Hydrogen**
Masses combining	81.8 g	$100 - 81.8 = 18.2$ g
Mass of one mole of each element	12 g	1 g
Number of moles combining	$\dfrac{81.8 \text{ g}}{12 \text{ g}} = 6.82$	$\dfrac{18.2 \text{ g}}{1 \text{ g}} = 18.2$
Simplest ratio of the number of moles	1 3	2.67 8

The empirical formula is C_3H_8

The relative formula mass of $C_3H_8 = (3 \times 12) + (8 \times 1)$
$$= 44$$

Thus the molecular formula is the same as the empirical formula

Since carbon atoms form four covalent bonds and hydrogen atoms form one, a possible structure is:

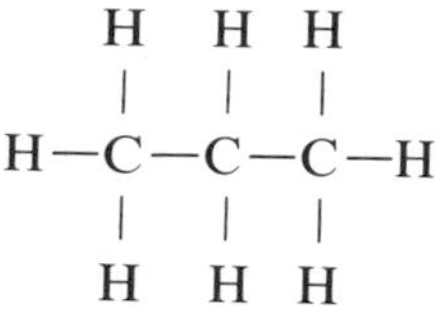

```
    H   H   H
    |   |   |
H — C — C — C — H
    |   |   |
    H   H   H
```

Questions

5 A compound used in antifreeze has the following analysis: 38.7 % C, 9.68 % H, with the remainder being oxygen. The relative molecular mass of the compound is 62. Calculate the empirical formula of the compound. Work out the molecular formula and suggest a possible structure.

6 When sulphur reacts with chlorine, the product consists of 47.4 % sulphur. The relative molecular mass of the compound is 135. Determine the empirical and molecular formulae of the compound.

7 Determine the empirical and molecular formulae of a hydrocarbon which consists of 82.8 % carbon, and whose relative molecular mass is estimated to be between 50 and 60.

2.3   The formulae of hydrated salts

There is water in the crystals of many salts. This water can be driven off by heating. For example, blue copper(II) sulphate crystals have the formula $CuSO_4 \cdot 5H_2O$. On heating, the crystals give off water vapour and turn to a white powder, which has the formula $CuSO_4$. This is called *anhydrous* copper(II) sulphate. The methods of calculation in this chapter can be used to work out the amount of water in crystals.

Example 2f

On heating a 5.40 g sample of hydrated iron(II) sulphate, $FeSO_4 \cdot nH_2O$, steam was driven off. The mass of the residue was 2.95 g. Calculate the value of n in the formula.

Answer	$FeSO_4$	H_2O
Masses combining	2.95 g	$5.40 - 2.95 = 2.45$ g
Mass of one mole	$(56 + 32 + 64) = 152$ g	18 g
Number of moles combining	$\dfrac{2.95\text{ g}}{152\text{ g}} = 0.0194$	$\dfrac{2.45\text{ g}}{18\text{ g}} = 0.136$
Simplest ratio	1	7

The value of n in the formula $FeSO_4 \cdot nH_2O$ is 7

Questions

8   Gypsum is a mineral that consists of hydrated calcium sulphate, $CaSO_4 \cdot$. The mineral contains 21.0 % water. Work out the formula of gypsum.

9   Rust consists of hydrated iron(III) oxide, $Fe_2O_3 \cdot nH_2O$. When a 5.50 g sample of rust is strongly heated, the residue has a mass of 4.71 g. Calculate the value of n in the formula.

10   Analysis of washing soda crystals gives the following results: 16.1 % Na, 4.2 % C, 16.8 % O, 62.9 % H_2O. Calculate the formula of the crystals.

2.4   Percentage composition

Once the formula of a compound is known, it can be used to work out the percentage composition. This information can show how much metal can be extracted from an ore. It can also be used to guide the people who formulate products such as fertilizers, medicines and washing powders.

Example 2g

Ammonium nitrate, NH_4NO_3, is commonly used in fertilizers to supply nitrogen to the soil. What is the percentage of nitrogen in ammonium nitrate?

Answer

The relative formula mass of ammonium nitrate $= 14 + 4 + 14 + (3 \times 16)$
$$= 80$$

The relative mass of nitrogen in the formula of the compound
$$= 2 \times 14$$
$$= 28$$

Thus the percentage of nitrogen $= \dfrac{28}{80} \times 100$

$$= 35\%$$

Questions

11 Calculate the percentage of nitrogen in the following compounds, which are used as fertilizers:

(a) ammonia, NH_3,

(b) calcium cyanamide, $CaCN_2$.

12 Calculate the percentage of copper in each of these minerals:

(a) cuprite, Cu_2O,

(b) malachite, $Cu_2(OH)_2CO_3$,

(c) bornite, Cu_5FeS_4.

Examination questions

1 During a chemistry lesson groups of pupils were burning strips of metal in air. The samples of metal were heated in crucibles with tightly fitting lids. During the heating the lids were lifted from time to time. Care was taken to prevent smoke escaping.

The mass of metal and the mass of solid oxide produced were recorded by each group. These results are shown in the table below.

Laboratory group	Mass of metal	Mass of solid oxide produced	Mass of oxygen combined
1	0.11 g	0.18 g	0.07 g
2	0.15 g	0.25 g	0.10 g
3	0.20 g	0.34 g	0.14 g
4	0.25 g	0.41 g	
5	0.30 g	0.50 g	

(a) Complete the final column in the table.

(b) Name the piece of apparatus most suitable for lifting the crucible lids during the heating.

(c) Why was it necessary to
  (i) lift the lids of the crucibles during the heating?
  (ii) prevent smoke escaping?

(d) Plot a graph of the mass of metal against mass of oxygen combined. Draw the best straight line graph through the origin.

(e) Suggest a reason why all the points on this graph do not fall exactly on a straight line.

(f) From the graph, find the mass of oxygen which would combine with 0.24 g of metal.

(g) What mass of oxygen would combine with 24 g of metal? (N.B. $0.24 \times 100 = 24$)

(h) The mass of 1 mole of atoms of the metal used is 24 g and the mass of 1 mole of oxygen atoms is 16 g. What is the simplest formula for the metal oxide? (Use M as the symbol for the metal.)

(i) In which periodic table group is the metal most likely to be placed?   **(EAEB)**

2 1.6 g of an oxide of iron contains 0.48 g of oxygen.

(a) How many grams of iron are present in the sample of oxide?

(b) How many grams of iron would combine with 48 g of oxygen to form this oxide?

(c) What is the formula of the oxide?   **(EAEB)**

3 Copper forms an oxide in which the ratio of the mass of copper to the mass of oxygen is 8 to 1. Find the formula for the oxide.   **(SUJB)**

4 1.42 g of an oxide of potassium contains 0.64 g of oxygen. Find the empirical (simplest) formula of this oxide.   **(WJEC)**

5 6.0 g of an oxide of lead gave 5.2 g of lead, Pb, on reduction by hydrogen. Calculate the empirical formula of the oxide.   **(JMB)**

6 (a) 15 g of an oxide A of vanadium(V) was found to contain 10.2 g of vanadium. Find the simplest (or empirical) formula of the oxide.

(b) On reaction with metallic vanadium, the oxide A formed a second oxide B which contained 16 g of oxygen combined with the molar mass of vanadium. Find the simplest formula of the oxide B and write an equation for its formation from the oxide A and vanadium.   **(WJEC)**

7 (a) 1.62 g of an oxide A of the element antimony (Sb) was found to contain 1.22 g of antimony. The molar mass of oxide A is 324 g/mol. Deduce the simplest (empirical) formula and the molecular formula of the oxide A.

(b) On heating strongly, the oxide A loses oxygen and forms a second oxide B which contains 32 g of oxygen combined with the molar mass of antimony. The molar mass of the oxide B is 308 g/mol.

(i) Deduce the simplest (empirical) formula and the molecular formula of oxide B.

(ii) Write an equation for the formation of oxide B from oxide A.   **(WJEC)**

8 Calculate the relative atomic mass of a metal M given that 0.65 g of the metal forms 1.61 g of a sulphate of formula MSO_4.   **(O)**

9 Find the empirical formula of a compound containing 2.44 % hydrogen, 39.02 % sulphur, 58.54 % oxygen.   **(SUJB)**

10 An oxide of lead contains 13.4 % of oxygen. What is the empirical formula of the oxide?   **(SUJB)**

11 The anhydrous chloride of a metal M of relative atomic mass 56.0 contains 44.1 % of M. Calculate the simplest formula of this compound.   **(O)**

12 A hydrocarbon X contains 82.76 % of carbon (by mass) and has a relative molecular mass of 58. X slowly reacts with chlorine in the presence of diffuse sunlight.

(a) Deduce the molecular formula of X.

(b) Give *two* pieces of evidence that suggest that X is an alkane.

(c) Define the term *isomerism*. Write down the full structural formulae of the isomers of X.   **(C)**

13 What is (a) the empirical formula, (b) the molecular formula of a compound

containing 4.04 % hydrogen, 24.24 % carbon and 71.72 % chlorine, and relative molar mass of 99?   **(SUJB)**

14 0.2 mole of a hydrocarbon contains 4.8 g of carbon and 0.8 g of hydrogen. Give
    (a) the relative molecular mass, (b) the molecular formula, (c) the name and structural formula.   **(O)**

15 The hydrocarbon naphthalene contains 6.25 per cent by mass of hydrogen. Its relative molecular mass (molecular weight) is 128. Deduce (showing your working)
    (a) the empirical formula, (b) the molecular formula of naphthalene.   **(O & C)**

16 A compound of nitrogen and hydrogen contains 87.5 % N and 12.5 % H.
    (a) Calculate the empirical formula of this compound. Given that its relative molecular mass is 32, what further deduction can you make?
    (b) Write down the structural formula for this compound to show clearly the arrangement of the atoms which make up the individual molecules.   **(O)**

17 4.92 g of magnesium sulphate crystals gave 2.40 g of anhydrous magnesium sulphate on heating to constant mass.
    (a) Calculate the mole ratio, $MgSO_4:H_2O$, in magnesium sulphate crystals.
    (b) Calculate the relative molecular mass of magnesium sulphate crystals.   **(JMB)**

18 A salt hydrate $S \cdot xH_2O$ of relative molar mass 322 contained 55.9 % of water of crystallization. Calculate the value of x.   **(SUJB)**

19 When 6.2 g of a hydrated sodium carbonate are heated gently until no further change in mass occurs, 5.3 g of anhydrous sodium carbonate remain. Calculate the formula of the hydrate.
    (Molar mass of anhydrous sodium carbonate = 106 g; molar mass of water = 18 g)   **(JMB)**

20 In an experiment to determine the formula of a hydrate of magnesium bromide $MgBr_2 \cdot xH_2O$, 7.30 g of the hydrate, on heating to constant mass, gave 4.60 g of the anhydrous salt. Calculate the value of x and hence deduce the formula of the hydrate.   **(WJEC)**

21 Calculate the percentage of sulphur in anhydrous sodium thiosulphate, $Na_2S_2O_3$.   **(SUJB)**

22 A manufacturer of fertilizer wishes to describe his product as 'containing 5.6 % N'. If his only source of nitrogen is carbamide (i.e. urea), NH_2CONH_2, calculate the mass of this substance which must be present in 100 tonnes of mixed fertilizer.   **(O)**

23 The potassium content of garden fertilizer is generally given as ' % by mass of K_2O' and ' % by mass of K'. If a particular fertilizer contains 4.7 % by mass of K_2O, what is the corresponding % by mass of K?   **(O)**

24 A sample of fertilizer containing ammonium sulphate as the only nitrogen-containing compound was found on analysis to contain 14 % nitrogen.
    (a) What is the percentage of nitrogen in pure ammonium sulphate?
    (b) Calculate the percentage of ammonium sulphate in the sample of fertilizer.
                                                                                 (WJEC)

25 (a) If 32.0 g of oxygen contains L molecules of oxygen gas, O_2, write down, in terms of L, the number of
        (i)   sulphur atoms, S, in 32.0 g of sulphur.
        (ii) sulphur dioxide molecules, SO_2, in 32.0 g of sulphur dioxide.
        (iii) sulphate ions, SO_4^{2-}, in 32.0 g of sulphate ions.
    (b) A crystal of urea, $CO(NH_2)_2$, has a mass of 0.6 g.
        (i)   Calculate the relative molecular mass of urea.
        (ii) How many moles of urea molecules are present in the 0.6 g crystal?
        (iii) L is known as the Avogadro constant and it has a numerical value of 6.0×10^{23}. How many molecules of urea are present in the crystal?   **(L)**

3 Masses of moles

An equation is a useful shorthand for reminding you what happens during a reaction. It can be used to calculate the correct proportions in which to mix the reactants. The expected yield of the products can also be calculated.

The rules for solving problems based on equations are shown in figure 3.1.

RULES

SOLVING PROBLEMS BASED ON EQUATIONS

1/ Write a balanced equation.

2/ In words, state what the equation tells you concerning the substances of interest.

3/ Change moles to masses.

4/ Scale the masses to those in the question.

Figure 3.1

Example 3a

In the thermit reaction, what mass of aluminium powder is needed to react with 8.0 g of iron(III) oxide?

Answer

Step 1 $2Al(s) + Fe_2O_3(s) \rightarrow Al_2O_3(s) + 2Fe(s)$

Step 2   Two moles of aluminium reacts with one mole of iron(III) oxide. (Note that in this question there is no need to mention the products in step 2.)

Step 3   The relative atomic mass of aluminium = 27

The relative formula mass of iron(III) oxide
$$= (2 \times 56) + (3 \times 16)$$
$$= 160$$

So $2 \times 27 \, g = 54 \, g$ of aluminium reacts with 160 g of iron(III) oxide

Step 4   The question refers to 8.0 g of iron(III) oxide

Let the mass of aluminium powder needed be x g

So $\dfrac{8.0}{160} = \dfrac{x}{54}$

$160x = 8.0 \times 54$ (cross multiplying)

$x = \dfrac{8.0 \times 54}{160}$ (dividing by 160)

$x = 2.7$

So 2.7 g of aluminium powder is needed

Example 3b

What mass of ethanol is formed when 4.5 g of glucose is fermented?

Answer

Step 1 $C_6H_{12}O_6(aq) \rightarrow 2C_2H_5OH(aq) + 2CO_2(g)$

Step 2   One mole of glucose is converted to two moles of ethanol. (The carbon dioxide can be ignored here because it is not mentioned in the question.)

Step 3   The relative molecular mass of glucose
$$= (6 \times 12) + (12 \times 1) + (6 \times 16)$$
$$= 180$$

The relative molecular mass of ethanol
$$= (2 \times 12) + (6 \times 1) + 16$$
$$= 46$$

Thus 180 g of glucose is converted to $2 \times 46 \, g$
$$= 92 \, g \text{ of ethanol}$$

Step 4   The question refers to 4.5 g of glucose

Let x g of ethanol be formed

So $\dfrac{4.5}{180} = \dfrac{x}{92}$

$180x = 4.5 \times 92$ (cross multiplying)

$x = \dfrac{4.5 \times 92}{180}$ (dividing by 180)

$x = 2.3$

So 2.3 g of ethanol is formed

Example 3c

What mass of magnesium sulphate crystals, $MgSO_4 \cdot 7H_2O$, can be made from 14.0 g of magnesium carbonate and an excess of dilute sulphuric acid?

Answer

Step 1   In solution, the equation is:

$$MgCO_3(s) + H_2SO_4(aq) \rightarrow MgSO_4(aq) + CO_2(g) + H_2O(l)$$

Step 2   One mole of magnesium carbonate reacts to form one mole of magnesium sulphate

Step 3   The relative formula mass of $MgCO_3$

$= 24 + 12 + (3 \times 16)$

$= 84$

The magnesium sulphate crystallizes as $MgSO_4 \cdot 7H_2O$

The relative formula mass of the crystals

$= 24 + 32 + (4 \times 16) + (7 \times 18)$

$= 246$

So, according to the equation, 246 g of crystals can be made from 84 g of magnesium carbonate

Step 4   The question refers to 14.0 g of magnesium carbonate

Let the mass of crystals formed be x g

So $\dfrac{14.0}{84} = \dfrac{x}{246}$

$84x = 14.0 \times 246$ (cross multiplying)

$x = \dfrac{14.0 \times 246}{84}$ (dividing by 84)

$x = 41$

So 41 g of crystals can be made

In the laboratory, masses are measured in grams. On an industrial scale, the unit of mass is the tonne. So long as all the masses are measured in the same units, the method of calculation is the same. If 2.7 g of aluminium is needed to react with 8.0 g of iron(III) oxide (see example 3a), then 2.7 tonnes of aluminium will be needed to react with 8.0 tonnes of the oxide. If 4.5 g of glucose ferments to give 2.3 g of ethanol (see example 3b), then 4.5 tonnes of glucose will ferment to give 2.3 tonnes of ethanol.

Questions

In questions **1** to **4**, you are given the equation, so step 1 has been done for you.

1 How much magnesium must be burned in oxygen to make 4.0 g of magnesium oxide?

$$2Mg(s) + O_2(g) \rightarrow 2MgO(s)$$

2 What mass of calcium oxide is formed when 25 g of calcium carbonate is decomposed by heat?

$$CaCO_3(s) \rightarrow CaO(s) + CO_2(g)$$

3 In the blast furnace, iron(III) oxide is reduced to iron by carbon monoxide:

$$Fe_2O_3(s) + 3CO(g) \rightarrow 2Fe(s) + 3CO_2(g)$$

(a) What mass of carbon monoxide is needed to reduce 16 tonnes of iron(III) oxide?

(b) What mass of iron is obtained from the reduction of 16 tonnes of the oxide?

4 What mass of coke is consumed in a blast furnace in the production of 2.8 tonnes of carbon monoxide?

$$2C(s) + O_2(g) \rightarrow 2CO(g)$$

In questions **5** to **8**, you are given the formulae but you have to write the balanced equation in step 1.

5 What is the loss in mass when 1.25 g of blue copper(II) sulphate crystals, $CuSO_4 \cdot 5H_2O$, is heated and decomposed to anhydrous copper(II) sulphate, $CuSO_4$?

6 What mass of ammonia, NH_3, is formed when 12 g of hydrogen, H_2, combines with nitrogen, N_2?

7 Lead(II) oxide, PbO, reacts with hydrogen to form lead and steam, H_2O. Calculate the mass of lead formed when 446 g of lead(II) oxide is reduced in this way.

8 What mass of sulphur is needed to react with 8.0 g copper to form copper(I) sulphide, Cu_2S?

In questions **9** to **14**, you are not given the equation or the formulae. Some of the formulae you will know; some you can work out, given a table of charges on common ions; some you will need to look up.

9 What mass of copper is formed when 4.0 g of copper(II) oxide is reduced by carbon?

10 What mass of sodium chloride can be crystallized from a solution formed by dissolving 5.0 g of sodium hydroxide in water and then neutralizing the alkali with dilute hydrochloric acid?

11 What mass of lead is displaced when 1.3 g of zinc is added to excess lead(II) nitrate?

12 What mass of aluminium is required to react with dry chlorine gas to make 4 g of anhydrous aluminium chloride?

13 What mass of sodium carbonate is formed when 8.4 g of sodium hydrogencarbonate is decomposed by heat? (The other products are water and carbon dioxide.)

14 What mass of 1,2-dibromoethane is formed when 5.6 g of ethene combines with bromine?

Examination questions

1 Magnesium sulphate can be obtained from magnesium oxide by the following reaction:

$$MgO + H_2SO_4 \rightarrow MgSO_4 + H_2O$$

(a) (i) What is the relative formula mass of magnesium oxide, MgO?
 (ii) What is the relative formula mass of magnesium sulphate, $MgSO_4$?

(b) What mass, in grams, of magnesium sulphate, $MgSO_4$, could be obtained from 1 g of magnesium oxide, MgO? **(EAEB)**

2 Carbon dioxide can be prepared commercially by roasting limestone in a kiln. The equation for the reaction is:

$$CaCO_3 \rightarrow CaO + CO_2$$

Given that the relative atomic masses are $Ca = 40$, $C = 12$ and $O = 16$, calculate the mass of carbon dioxide and calcium oxide produced from 50 tonnes of limestone. **(WJEC)**

3 The equation for the reaction between ammonia and heated copper(II) oxide is:

$$3CuO + 2NH_3 = 3Cu + N_2 + 3H_2O$$

Calculate the mass of copper which could be produced by the reaction of 10 g of ammonia with an excess of copper(II) oxide. **(NISEC)**

4 (a) Rewrite balanced forms of the following equations, which refer to the extraction of copper from its ore chalcopyrite ($CuFeS_2$):

$$CuFeS_2 + O_2 = Cu_2S + FeO + SO_2$$
$$Cu_2S + O_2 = Cu_2O + SO_2$$
$$Cu_2S + Cu_2O = Cu + SO_2$$

(b) Calculate the maximum amount of copper which can be obtained from 100 kg of chalcopyrite. **(O)**

5   (a) Calculate the maximum mass of iron which could be produced from a blast
         furnace charge of 80 tonnes of haematite ore, Fe_2O_3.
    (b) The mass of iron actually extracted amounted only to 45 tonnes. Give *two*
         reasons to account for this low yield.   **(O)**

6   (a) (i)   Describe how you would prepare dry crystals of sodium sulphate
                 $(Na_2SO_4 \cdot 10H_2O)$, starting from dilute sulphuric acid and dilute aqueous
                 sodium hydroxide.
         (ii) What mass (in grams) of the *hydrated* crystals would you expect to obtain
                 from 0.1 mole of sodium hydroxide?
    (b) Excess aqueous barium chloride was added to a solution of 0.025 mole of
         sodium sulphate dissolved in water. Write down the ionic equation for the
         reaction and calculate the mass of the dry product.   **(O & C)**

7   What will be the mass of the residue after heating 10.0 g of crystalline barium
    chloride, $BaCl_2 \cdot 2H_2O$, until all the water of crystallization has been driven off?
                                                                         (O & C)

8   Each year 200 million tonnes of new carbon monoxide is released into the
    atmosphere. It is thought that almost all of the CO is eventually oxidized to carbon
    dioxide.
    (a) Write down an equation to represent the oxidation reaction.
    (b) Calculate the mass of carbon dioxide produced yearly by this transformation.
                                                                              (O)

9   Most sulphur dioxide in the atmosphere is oxidized to sulphur trioxide which is
    then converted to sulphuric acid by reaction with rain water. The overall reaction
    may be represented by:

$$2SO_2 + O_2 + 2H_2O = 2H_2SO_4$$

    Calculate the daily production of sulphuric acid from a smelting plant which in that
    period discharges 320 tonnes of SO_2 into the atmosphere.   **(O)**

10 The use of limestone has been suggested as a means of reducing SO_2 pollution from
    power stations. It would be added in powder form to the coal:

$$CaCO_3 \xrightarrow{\text{heat}} CaO + CO_2 \quad \text{then} \quad CaO + SO_2 \longrightarrow CaSO_3$$

    (a) Calculate the mass of limestone which in theory would be needed to deal with
         an output of 320 tonnes of sulphur dioxide.
    (b) Give *one* reason why in practice much more limestone would be needed to deal
         with all the SO_2 than indicated by your answer to this calculation.   **(O)**

11 Proposals have been made to eliminate the pollutant sulphur dioxide by passing the
    waste gases from a power station into the sea, whereupon the dissolved oxygen
    would oxidize the SO_2 to the less harmful sulphate ion (SO_4^{2-}) by means of the
    reaction

$$2SO_2 + O_2 + 4OH^- = 2SO_4^{2-} + 2H_2O$$

    Calculate the number of tonnes of dissolved oxygen which would be consumed by
    480 tonnes of SO_2, a typical daily output from a large power station.   **(O)**

12 An important industrial reaction is the manufacture of ethanol from ethene
    (ethylene). It is the reverse of the dehydration reaction sometimes used to prepare
    ethene from ethanol.
    (a) Write down a balanced equation to represent the hydration of ethene to form
         ethanol.
    (b) Calculate the maximum mass of ethanol which could be prepared from 280 kg
         of ethene.

(c)   If the reaction only yielded 400 kg of pure ethanol, calculate the percentage
yield for the reaction.   (O)

13   Ethyl ethanoate ($CH_3CO_2C_2H_5$) can be prepared from ethanol and ethanoic acid.
(a)   Write the equation for the reaction.
(b)   Calculate the theoretical yield of ethyl ethanoate which can be obtained from
23 g of ethanol.
(c)   Experimentally it was found that 33 g of ethyl ethanoate were obtained from
23 g of ethanol. Calculate the percentage yield.   (C)

14   DDT (I) may be converted into an ethene-related compound DDE (II) by the
action of micro-organisms in soil or river water.

(I)   (II)

Relative molecular   Relative molecular
mass = 354.5   mass = 318

(a)   A farmer sprays 0.71 kg of DDT on his fields. What mass of DDE will be
produced from this amount of insecticide?
(b)   What mass of DDE would in fact be produced if the conversion is only 10 %
complete?   (O)

4 Moles of gases

4.1   Gas volumes

Many chemical reactions involve gases. It is easier to measure the volume of a gas than its mass. Fortunately, it is also easy to work out the number of moles of a gas from its volume, provided that the measurements are made at a known temperature and pressure.

The volume of a sample of a gas depends on three things:

✳ the temperature, T
✳ the pressure, P
✳ the number of moles of molecules, n.

Various units of volume are used in everyday life and in the laboratory. Petrol is sold in litres (l). Smaller volumes may be measured in millilitres (ml). There are 1000 ml in one litre. A millilitre is the same as a cubic centimetre (cm^3). Thus there are also 1000 cm^3 in one litre.

$$10\,cm = 1\,dm\ \text{(decimetre)}$$

So $\quad 10\,cm \times 10\,cm \times 10\,cm = 1000\,cm^3$
$$= 1\,dm \times 1\,dm \times 1\,dm$$
$$= 1\,dm^3$$
$$= 1\,litre$$

To summarize $\quad 1\,cm^3 = 1\,ml$
and $\qquad\qquad 1\,dm^3 = 1\,litre$

The effect of temperature on gas volumes

If the pressure and the amount of gas are kept constant, the effect of changes in temperature on the volume may be studied. Gases expand when the temperature rises. They expand in a regular way, as shown by figure 4.1.

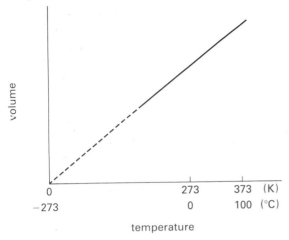

Figure 4.1   How the volume of a gas changes with temperature

Normally in the laboratory, temperatures are measured on the Celsius scale. On this scale, water freezes at 0 °C and boils at 100 °C. When studying gases, temperatures must be converted to the Kelvin scale. On this scale, water freezes at 273 K and boils at 373 K. Absolute zero is -273 °C, or 0 K.

As figure 4.1 shows, the volume of a gas is proportional to the temperature on the Kelvin scale. (One variable is proportional to another when the graph is a straight line through the origin.) If the temperature is doubled, the volume is doubled. This is an example of Charles's law, which states that the volume, V, of a fixed amount of gas is proportional to the temperature on the Kelvin scale, T, if the pressure is kept constant, i.e.

$*\ V \propto T$

Example 4a
If the volume of a sample of gas is 25 cm^3 at 20 °C, what will its volume be at 137 °C, if the pressure remains constant?

Answer
On the Kelvin scale, the starting temperature
$$= 273 + 20$$
$$= 293 \, K$$

The final temperature $= 273 + 137$
$$= 410 \, K$$

The volume increases in proportion to the temperature on the Kelvin scale

The new volume $\quad = 25 \, cm^3 \times \dfrac{410 \, K}{293 \, K}$

$$= 35 \, cm^3$$

Questions

1   Work out the number of cubic centimetres in each of the following volumes: (a) 2 l,   (b) 50 ml,   (c) 0.5 dm³,   (d) one gallon, which is 4.55 litres,   (e) one pint, which is 568 ml.
     Write the answers as a number followed by the unit cm³.

2   Work out the number of cubic decimetres (dm³) in each of the following volumes: (a) 5000 cm³,   (b) 2.5 l,   (c) 500 ml,   (d) 1 m³.
     Write the answers as a number followed by the unit dm³.

3   Convert these temperatures from the Celsius scale to the Kelvin scale:
     (a) the boiling point of sulphur, 445 °C,
     (b) the melting point of iron, 1540 °C,
     (c) the boiling point of oxygen, −183 °C,
     (d) the melting point of bromine, −7 °C.

4   In this question, assume that the pressure remains constant.
     (a) A sample of hydrogen occupies 22.4 dm³ at 273 K. What is its volume at 293 K?
     (b) The volume of a sample of nitrogen is 40 cm³ at 298 K. What is its volume at 373 K?
     (c) The volume of a sample of carbon dioxide is 100 cm³ at 300 K. What is its volume at 200 K?
     (d) If the volume of some ammonia is 60 cm³ at 80 °C, what is its volume at 10 °C?
     (e) If the volume of some sulphur dioxide is 30 cm³ at 15 °C, what is its volume at 95 °C?

The effect of pressure on gas volumes

To investigate the effect of pressure on the volume of a gas, the temperature must be kept fixed and so must the amount of gas. As the pressure goes up, the volume goes down. If the pressure is doubled, the volume is halved. This is Boyle's law.

✱ $P \times V = \text{constant}$

If the volume of a gas sample is V_1 at pressure P_1 and V_2 at pressure P_2, then

$$P_1 \times V_1 = P_2 \times V_2$$

Pressure is a measure of force per unit area. The units of force are newtons (N) and the units of area are square metres (m²). So the unit of pressure is N/m². This unit is sometimes called the pascal (Pa). Often, gas pressures are compared with standard atmospheric pressure.

$$\begin{aligned} \text{The pressure of one atmosphere} &= 101\,325\ \text{N/m}^2 \\ &= 101\,325\ \text{Pa} \\ &= 101.325\ \text{kPa} \end{aligned}$$

In the laboratory, pressure is often measured using a mercury barometer. The pressure is measured by the height of the mercury column. A pressure of one atmosphere will support a mercury column 760 mm high. The chemical symbol for mercury is Hg. A pressure of one atmosphere can be written 760 mmHg.

Example 4b
If the volume of a sample of gas is $80 \, cm^3$ at $20 \, °C$ and at a pressure of $760 \, mmHg$, what will be the volume if the pressure is increased to $800 \, mmHg$?

Answer
In this question $P_1 = 760 \, mmHg$ and $V_1 = 80 \, cm^3$
                            $P_2 = 800 \, mmHg$ and V_2 is to be calculated

Therefore $760 \, mmHg \times 80 \, cm^3 = 800 \, mmHg \times V_2 \, cm^3$

So $$V_2 = \frac{760 \, mmHg}{800 \, mmHg} \times 80 \, cm^3$$

                            $$= 76 \, cm^3$$

As expected, the volume decreases when the pressure is raised

Questions
5   In this question, assume that the temperature remains constant.
   (a) If the volume of a sample of hydrogen is $100 \, cm^3$ at one atmosphere pressure, what will be the volume if the pressure is increased to two atmospheres?
   (b) If the volume of some oxygen is $50 \, cm^3$ at a pressure of $100 \, N/m^2$, what will be the volume if the pressure is lowered to $80 \, N/m^2$?
   (c) If the volume of some carbon dioxide is $750 \, cm^3$ at a pressure of $760 \, mmHg$, what will be the volume when the pressure is $780 \, mmHg$?
   (d) If the volume of some nitrogen is $120 \, cm^3$ at $730 \, mmHg$, what will be its volume when the pressure is $780 \, mmHg$?
6   A gas has a volume of $500 \, cm^3$ at a pressure of one atmosphere. If the temperature is unchanged, what will be its volume at the following pressures: (a) five atmospheres,   (b) $110 \, N/m^2$,   (c) $740 \, mmHg$?

The effect of the amount of gas on its volume
If the pressure and temperature are fixed, then the volume of gas depends on the number of gas molecules present. The type of gas does not matter. This is Avogadro's law. The volume of a gas, V, is proportional to the amount of gas, n, if the pressure and temperature are kept constant. Amounts are measured in moles.
   Two sets of conditions are often used to compare one gas with another.

✱ Standard temperature and pressure, s.t.p.

The standard conditions for gases are $273 \, K$ ($0 \, °C$) and one atmosphere pressure. Under these conditions, one mole of any gas has a volume of $22.4 \, dm^3$ ($22 \, 400 \, cm^3$).

✱ Room temperature ($20 \, °C$) and pressure (one atmosphere)

Most experiments in the laboratory are done under these conditions. The volume of one mole of gas is larger at room temperature than at s.t.p. At room temperature and pressure, the volume of one mole of any gas is $24\,dm^3$ $(24\,000\,cm^3)$.

The following formulae can be used in calculations involving gases:

* Number of moles of gas $= \dfrac{\text{Volume of gas } (cm^3)}{\text{Volume of one mole of gas } (cm^3)}$

* Volume of gas (cm^3) = Number of moles × Volume of one mole (cm^3)

Questions

7 How many moles of gas are there, at room temperature, in
 (a) $24\,000\,cm^3$ of carbon dioxide?
 (b) $48\,cm^3$ of hydrogen?
 (c) $240\,000\,cm^3$ of chlorine?
 (d) $3\,dm^3$ of ammonia?
 (e) $72\,dm^3$ of oxygen?

8 How many moles of gas are there, at s.t.p. in
 (a) $22\,400\,cm^3$ of helium?
 (b) $89.6\,cm^3$ of argon?
 (c) $224\,dm^3$ of carbon monoxide?
 (d) $0.112\,dm^3$ of hydrogen sulphide?

9 Work out the volumes of the following amounts of gas (i) at s.t.p. and (ii) at room temperature and pressure:
 (a) $2\,mol$ of nitrogen,
 (b) $10\,mol$ of hydrogen chloride,
 (c) $0.01\,mol$ of neon,
 (d) $0.000\,002\,mol$ of carbon dioxide,
 (e) $0.125\,mol$ of helium.

10 Work out the volumes of the following quantities of gas (i) at s.t.p. and (ii) at room temperature and pressure:
 (a) $2\,g$ of hydrogen, H_2,
 (b) $3.2\,g$ of oxygen gas, O_2,
 (c) $0.0028\,g$ of nitrogen gas, N_2,
 (d) $0.011\,g$ of carbon dioxide, CO_2,
 (e) $0.000\,016\,g$ of methane, CH_4.

4.2 Relative molecular masses

Measurements of gas density can be used to measure the relative molecular mass of a gas.

At s.t.p., the volume of one mole of a gas is $22.4\,dm^3$. If the density of a gas is measured under these conditions, the mass of $22.4\,dm^3$ can be calculated, and this is the mass of one mole.

At room temperature and pressure, the volume of one mole of a gas is $24\,dm^3$. If the density of a gas is measured under these conditions, the mass of

$24\,dm^3$ can be calculated, and this is the mass of one mole.

✱ Mass of gas (g) = Volume of gas (dm^3) × Density of gas (g/dm^3)

Example 4c
The density of an alkaline gas is $0.71\,g/dm^3$ at room temperature and pressure. What is the mass of one mole of the gas?

Answer
The volume of one mole of the gas is $24\,dm^3$ at room temperature and pressure
The mass of $24\,dm^3$ of the gas $= 24\,dm^3 \times 0.71\,g/dm^3$
$= 17\,g$

Therefore, the mass of one mole of the gas $= 17\,g$
The relative molecular mass of this alkaline gas $= 17$

Questions
11 One litre of a gas has a mass of $2.32\,g$ at s.t.p. Calculate the mass of one mole of the gas.
12 One litre ($1\,dm^3$) of a gas has a mass of $1.58\,g$ at room temperature and pressure. Calculate the mass of one mole of the gas.
13 $30\,cm^3$ of a gaseous oxide of nitrogen reacts with hot iron to form $30\,cm^3$ of nitrogen, N_2. The mass of $1\,dm^3$ of the oxide is $1.83\,g$ at room temperature and pressure.
   (a) What happens to the oxygen in the gaseous oxide when it reacts with iron?
   (b) How many moles of nitrogen are formed when one mole of the gaseous oxide reacts with iron?
   (c) How many atoms of nitrogen must there be in each molecule of the gaseous oxide?
   (d) Work out the mass of one mole of the gaseous oxide.
   (e) Use your answers to (c) and (d) to determine the molecular formula of the gaseous oxide.
   (f) Write an equation for the reaction of the gaseous oxide with iron. Include state symbols and assume that iron(III) oxide is formed. The oxide ion is O^{2-}.
14 The density of a gaseous oxide of carbon is $1.15\,g/dm^3$ at room temperature and pressure. Work out the mass of one mole of the gas and give its formula.
15 A hydrocarbon consists of 80% carbon and 20% hydrogen. The density of the hydrocarbon, which is a gas, is $1.34\,g/dm^3$ at s.t.p.
   (a) Work out the simplest (empirical) formula of the hydrocarbon.
   (b) Work out the relative molecular mass of the hydrocarbon.
   (c) What is the molecular formula of the hydrocarbon?

4.3   Calculations from equations

Calculations from equations for reactions which involve gases are carried out in a similar way to mass calculations. Gas volume calculations are easier because the volume of a gas depends only on the number of moles. It does not matter which gas is involved.

RULES

GASES IN EQUATIONS

1/ Write a balanced equation.

2/ In words, state what the equation tells you about the substances of interest.

3/ Change moles into masses or volumes (if gases).

4/ Scale the masses or volumes to those in the question.

N.B. 1 mol of any gas has a volume of 24 000 cm³ at room temperature and pressure (or 22 400 cm³ at s.t.p.).

Figure 4.2

Example 4d

What volume of hydrogen, measured at s.t.p., is formed when 0.35 g of lithium reacts with water?

Answer

Step 1 $2Li(s) + 2H_2O(l) \rightarrow 2LiOH(aq) + H_2(g)$

Step 2   Two moles of lithium reacts to form one mole of hydrogen molecules

Step 3   The relative atomic mass of lithium = 7

So the mass of 2 mol of lithium   = 14 g

At s.t.p. the volume of 1 mol of hydrogen = 22.4 dm³

So the equation shows that 14 g of lithium produces 22 400 cm³ of hydrogen

Step 4   The question refers to 0.35 g of lithium

This is $\dfrac{0.35\,g}{14\,g} = 0.025$ of the amount in the equation

So the amount of hydrogen formed $= 0.025 \times 22\,400\,cm^3$
$$= 560\,cm^3$$

Questions

16   Calculate the volume of carbon dioxide given off, at s.t.p., when 0.9 g of glucose ferments:

$$C_6H_{12}O_6(aq) \rightarrow 2C_2H_5OH(aq) + 2CO_2(g)$$

17   Carbon dioxide is often made by the action of hydrochloric acid on marble chips (calcium carbonate):

$$CaCO_3(s) + 2HCl(aq) \rightarrow CaCl_2(aq) + CO_2(g) + H_2O(l)$$

(a) If the volume of a gas jar is approximately $500\,cm^3$, what volume of carbon dioxide is needed to fill six gas jars, at room temperature and pressure?
(b) How many moles of carbon dioxide are needed to fill six gas jars?
(c) How many moles of calcium carbonate must react to give the number of moles of carbon dioxide required in (b)?
(d) What mass of calcium carbonate must react to fill six gas jars?

18   What volume of hydrogen is given off, at room temperature and pressure, when 0.65 g of zinc reacts with excess sulphuric acid?

19   What volume of oxygen, measured at s.t.p., is given off when 0.76 g of potassium nitrate is decomposed by heat according to the equation

$$2KNO_3(s) \rightarrow 2KNO_2(s) + O_2(g)$$

Volume ratios

Gas volume calculations are especially easy when the reactants and products involved are all gases.

Equal volumes of gases contain equal numbers of moles (if measured under the same conditions). This means that the ratio of the gas volumes in a reaction must be the same as the ratio of the numbers of moles in the equation.

Consider the decomposition of ammonia, which occurs when the gas is passed over very hot iron:

$$2NH_3(g) \rightarrow N_2(g) + 3H_2(g)$$
  2 mol   1 mol   3 mol

If $20\,cm^3$ of ammonia is decomposed, then $10\,cm^3$ of nitrogen and $30\,cm^3$ of hydrogen will be formed.

Example 4e

What volume of oxygen is needed to react with $40 \, cm^3$ of methane, and what volume of carbon dioxide is formed, if all gas volumes are measured under the same conditions?

Answer

The equation for the reaction is:

$$CH_4(g) + 2O_2(g) \rightarrow CO_2(g) + 2H_2O(l)$$

1 mol   2 mol   1 mol

So $40 \, cm^3$ of methane will react with $80 \, cm^3$ of oxygen to form $40 \, cm^3$ of carbon dioxide

The ratio of the volumes is the same as the ratio of the numbers of moles

Questions

20   (a) What volume of oxygen contains the same number of molecules as $100 \, cm^3$ of nitrogen?

     (b) What volume of argon contains the same number of atoms as $2 \, cm^3$ of helium?

     (c) What volume of chlorine contains the same number of molecules as $103 \, dm^3$ of ethane?

21   What volume of oxygen is needed to react with $50 \, cm^3$ ethane, $C_2H_6(g)$, when it burns, and what volume of carbon dioxide is formed?

22   Nitrogen oxide, NO, reacts with oxygen, O_2, to give nitrogen dioxide, NO_2. Starting with $50 \, cm^3$ of nitrogen oxide, what volume of oxygen will be needed to react with it, and what volume of nitrogen dioxide will be formed, if all volumes are measured under the same conditions?

23   When $100 \, cm^3$ of hydrogen bromide reacts with $80 \, cm^3$ of ammonia, a white solid forms and there is an excess of one gas.

$$NH_3(g) + HBr(g) \rightarrow NH_4Br(s)$$

Which gas is in excess, and what volume of this gas remains unreacted?

24   The following reaction takes place when chlorine gas is shaken with excess ammonia solution:

$$3Cl_2(g) + 8NH_3(aq) \rightarrow N_2(g) + 6NH_4Cl(aq)$$

What volume of nitrogen is formed when $72 \, cm^3$ of chlorine is shaken with excess ammonia solution?

Analysis of hydrocarbons

Measurements of gas volumes can be used to find the formulae of hydrocarbon gases. The hydrocarbon is burned in an excess of oxygen. After reaction, the volume of carbon dioxide formed is measured, and the volume of excess oxygen is determined.

Example 4f
20 cm^3 of a hydrocarbon gas requires 100 cm^3 of oxygen for combustion, and produces 60 cm^3 of carbon dioxide. What is the formula of the hydrocarbon?

Answer
The gas is a hydrocarbon, so it consists of carbon and hydrogen only. Its formula can be written C_xH_y. The problem is to find the values of x and y. The volume ratios are

20 cm^3 C_xH_y : 100 cm^3 O_2 : 60 cm^3 CO_2

This can be simplified to these volume ratios, by dividing by 20:

1 volume C_xH_y : 5 volumes O_2 : 3 volumes CO_2

But equal volumes of gases contain equal numbers of moles of gas, so the mole ratios must be the same:

1 mol C_xH_y : 5 mol O_2 : 3 mol CO_2

Thus the equation can be written:

$C_xH_y(g) + 5O_2(g) \rightarrow 3CO_2(g)$ + Water (which condenses)

All the carbon in the carbon dioxide must have come from the hydrocarbon

The value of x must be 3, if the equation is to balance

Three of the five moles of oxygen molecules are used to make carbon dioxide, which leaves two to combine with the hydrogen to form water. Two moles of oxygen, $2O_2$, will form four moles of water, $4H_2O$. So the value of y is 8

The formula of the hydrocarbon is C_3H_8

Questions
25 20 cm^3 of a hydrocarbon gas was exploded with 100 cm^3 of oxygen. After the reaction, 40 cm^3 of carbon dioxide was formed and 30 cm^3 of excess oxygen remained. What is the formula of the hydrocarbon?
26 10 cm^3 of a hydrocarbon gas reacts with 90 cm^3 of oxygen to produce 60 cm^3 of carbon dioxide. What is the formula of the hydrocarbon?

Examination questions

1 The element chlorine exists in two isotopic forms which may be represented by:

Form A: $^{35}_{17}Cl$ Form B: $^{37}_{17}Cl$

The upper figure is the mass number and the lower figure is the atomic number in each case.

(a) Complete the following list:

Number of protons in   A............   B............
Number of electrons in   A............   B............
Number of neutrons in   A............   B............

(b) In natural chlorine, the proportion by mass of A to B is 3:1. Use this information to calculate the relative atomic mass of chlorine. Show clearly how you have arrived at your answer.

(c) What would the volume of 240 cm³ of chlorine gas become if the temperature remained constant and the pressure on it was doubled?

(d) If room temperature was 20 °C, at what temperature would the volume of the chlorine be doubled, given that the pressure remained constant?

(e) 240 cm³ of fluorine gas, measured at room temperature and atmospheric pressure, has a mass of 0.38 g. Use this information to work out the atomicity of fluorine. Show clearly how you have arrived at your answer.   **(L)**

2 Five students investigating the thermal decomposition of a compound found that a gas was evolved. They each then repeated their experiments, finding the loss in mass of the compound and measuring the volume of gas evolved at room temperature and pressure. Their results are as follows:

Student	A	B	C	D	E
Loss in mass (g)	0.060	0.032	0.107	0.083	0.090
Volume of gas (cm³)	45	24	80	62	75

(a) Plot the results as a graph.

(b) (i) One student made an error in measuring the volume of his gas. Draw a ring round this point.

(ii) Draw a straight line through the other points.

(c) Use your graph to find the mass of 100 cm³ of gas.

Below are the densities of some common gases at room temperature and pressure.

Gas	Density (g litre⁻¹)
Hydrogen	0.008
Nitrogen	1.16
Oxygen	1.33
Carbon dioxide	1.83

(d) State, with a reason, which gas might have been the one evolved during the experiments.

(e) Sketch an apparatus in which this experiment might have been carried out.

(JMB)

3 An experiment was carried out to investigate how the times taken for 100 cm³ of different gases to diffuse depended on their relative molecular masses (M_r). The graph opposite was obtained.

(a) Under the conditions of the experiment, how long would it take for 100 cm³ of each of the following gases to diffuse?

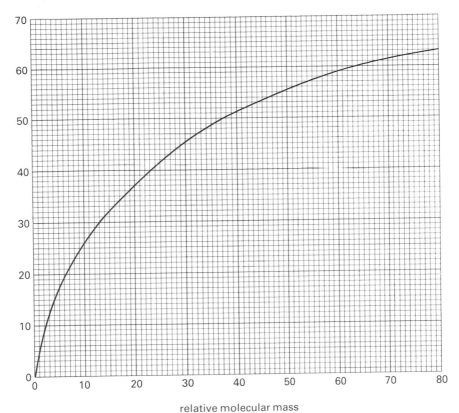

relative molecular mass

Gas	M_r	Time (seconds)
Neon	20	
Oxygen	32	
Sulphur dioxide	64	

(b) Write down the formula of the alkane which would take 58 seconds for $100 \, cm^3$ to diffuse.

(c) Suggest the identity of two gases, both of which contain carbon and would take 44 seconds for $100 \, cm^3$ to diffuse.   (C)

4 0.1 g of a mixture of magnesium and magnesium oxide when treated with an excess of aqueous hydrochloric acid gave $60 \, cm^3$ of dry hydrogen at room temperature and pressure.

(a) Write an equation for the reaction between magnesium and hydrochloric acid.

(b) Calculate the number of moles of hydrogen, H_2, formed in the reaction.

(c) Using the result from (b) and the equation given in (a), calculate the number of moles of magnesium, Mg, that would be required to give this amount of hydrogen.

(d) What mass of magnesium does this represent?

(e) Calculate the percentage of magnesium in the mixture.

(f) If the hydrogen had been measured at $0\,°C$ at the same pressure, would the volume of hydrogen have been larger or smaller than $60 \, cm^3$?   (WJEC)

5    $2Li(s) + 2H_2O(l) \rightarrow 2LiOH(aq) + H_2(g)$

From the equation, it can be deduced that 2 moles of lithium atoms would react with an excess of water to give 1 mole of hydrogen molecules. The following experiment was done in order to verify this statement.

A small pellet of lithium was taken from a stock bottle and washed in an organic solvent to remove the oil in which it had been stored. The pellet was quickly dried between filter papers and weighed. The dry pellet was then reacted with water and the volume of hydrogen was collected and measured at room temperature and pressure.

Mass of lithium $= 0.021\,g$
Volume of hydrogen $= 35\,cm^3$

(a) Draw a diagram of the assembled apparatus which could have been used for this experiment.
(b) Use the results obtained to verify that 2 moles of lithium atoms displace 1 mole of hydrogen molecules.
(c) This experimental value for the volume of one mole of hydrogen is slightly lower than the expected value. What do you think is the chief source of error in the experiment?
                                                                                (JMB)

6   0.500 g of an alloy of iron, carbon and silicon (silicon steel) when reacted with an excess of hydrochloric acid gave 192 cm³ of hydrogen at room temperature and pressure. Calculate the percentage of iron in silicon steel.   (C)

7   Calculate the volume occupied by 11 g of carbon dioxide (a) at s.t.p. and (b) at 546 °C and 2 atmospheres pressure.   (O)

8   A store used for keeping flammable materials measures 50 dm × 44.8 dm × 20 dm. It is suggested that it is fitted with a cylinder of liquefied carbon dioxide which could be rapidly released (as gaseous CO_2) if the contents of the store catch fire.
(a) Calculate the volume (in dm³) of the store.
(b) Calculate the mass of carbon dioxide which would be required to fill the store in the event of fire risk. (Assume all measurements are made at s.t.p.)   (O)

9   The equation for the thermal decomposition of sodium hydrogencarbonate is:

$2NaHCO_3(s) = Na_2CO_3(s) + H_2O(g) + CO_2(g)$

(a) Explain what is meant by *thermal decomposition*.
(b) If 21 g of sodium hydrogencarbonate were decomposed, calculate
     (i) the mass of the residue formed.
     (ii) the volume of carbon dioxide evolved, measured at s.t.p.   (C)

10   Lithium hydride has been used to generate hydrogen, since when it is treated with water, hydrogen is evolved according to the equation

$LiH + H_2O = LiOH + H_2$

Calculate the volume of hydrogen, in dm³ at s.t.p., which could be obtained in this way from one kilogram of lithium hydride.   (O & C)

11   In the first stage of the manufacture of nitric acid from ammonia, the ammonia is oxidized to nitrogen monoxide according to the equation

$4NH_3 + 5O_2 = 4NO + 6H_2O$

Calculate the volume of oxygen (in dm³ at s.t.p.) needed to oxidize one mole of ammonia in this way, and the volume of nitrogen monoxide thereby produced.
                                                                                (O & C)

12   By complete decomposition of the hydrogen peroxide in a certain aqueous solution of this substance, $1.40 \, dm^3$ of oxygen (measured at s.t.p.) were obtained. What mass of hydrogen peroxide did this solution contain?   **(O & C)**

13   What volume of oxygen is required to burn completely 1 mole of ethene, and what extra volume is required for 1 mole of ethane? (All volumes measured at s.t.p.)
    (SUJB)

14   Hydrogen sulphide can be prepared by treating iron(II) sulphide with dilute sulphuric acid. What volume of hydrogen sulphide, measured at s.t.p., can be obtained from $17.6 \, g$ of iron(II) sulphide?   **(SUJB)**

15   $200 \, cm^3$ of a mixture of methane and ethene at s.t.p. were bubbled through water saturated with bromine. $0.8 \, g$ of bromine was found to have reacted. What was the composition of the original mixture?   **(SUJB)**

16   Calcium nitrate decomposes on heating according to the equation

$$2Ca(NO_3)_2 = 2CaO + 4NO_2 + O_2$$

The relative molecular mass of calcium nitrate is 164.

    What volume at s.t.p. of (a) nitrogen(IV) oxide (nitrogen dioxide), and (b) oxygen is evolved when $16.4 \, g$ of calcium nitrate is heated to constant weight?   **(SUJB)**

17   For the reaction

$$Pb_3O_4(s) + 4H_2(g) = 3Pb(s) + 4H_2O(g)$$

what volume of hydrogen, measured at standard temperature and pressure, is required to react with $5 \, g$ of Pb_3O_4?   **(NISEC)**

18   The manure produced by one cow per day can be made to yield, by fermentation, $1.3 \times 10^3 \, dm^3$ of biogas (measured at s.t.p.). Biogas contains 55% by volume of methane, 44% of carbon dioxide and traces of other substances including hydrogen sulphide. Calculate, at s.t.p.
    (a)   the volume of methane indirectly produced by one cow per day.
    (b)   the volume of carbon dioxide produced by burning this volume of methane (using the equation for the complete combustion of methane to help you).
    (c)   the total volume of carbon dioxide remaining after the burning of 1 cow-day of biogas.   **(O)**

19   Recent research shows that a 'typical' fire from $1 \, dm^3$ of spilt petrol needs $560 \, dm^3$ of gaseous carbon dioxide for it to be extinguished. Calculate the maximum volume of burning petrol which can be safely put out by an appliance containing $2.2 \, kg$ of liquid CO_2. (All gaseous volumes are measured at s.t.p.)   **(O)**

20   Calculate the volume of ethene (measured at s.t.p.) required to produce $112 \, g$ of polyethene ('Polythene'), assuming that 100% conversion to polymer occurs.   **(O)**

21   Metallic iron dissolves in dilute hydrochloric acid according to the equation

$$Fe(s) + 2HCl(aq) = FeCl_2(aq) + H_2(g)$$

A sample of iron powder, known to contain no significant proportion of any other metal, was found to have partially corroded. To determine how much iron was still present as the metal, $1.25 \, g$ of the sample was treated with excess hydrochloric acid until reaction ceased. $450 \, cm^3$ of hydrogen (measured at s.t.p.) were found to have been evolved. Calculate the number of grams of metallic iron present in $100 \, g$ of the sample.   **(O & C)**

22   $0.2 \, mol$ of a hydrocarbon T contains $4.8 \, g$ of carbon and $0.8 \, g$ of hydrogen. Calculate (a) the relative molecular mass, (b) the molecular formula and (c) the volume of carbon dioxide formed when $100 \, cm^3$ of T were completely burnt in oxygen. (The volumes were measured at the same temperature and pressure.)   **(C)**

23   A compound X contains 12.8 % carbon, 2.1 % hydrogen and 85.1 % bromine. 56cm^3 at s.t.p. of the vapour of X had a mass of 0.47 g. Calculate (a) the empirical formula and (b) the relative molar mass.
   (c)   State the molecular formula of X, draw its structural formula, given that it is symmetrical, and suggest how X may be prepared.   **(SUJB)**

24   A liquid hydrocarbon was analysed and it was found that 1.72 g of the hydrocarbon contained 1.44 g of carbon.
   (a)   Deduce the empirical (simplest) formula of the hydrocarbon.
   (b) 1dm^3 (litre) of the hydrocarbon vapour at room temperature and pressure was found to have a mass of 3.583 g ($3\frac{7}{12}$ g).
       (i)   Calculate the molar mass of the hydrocarbon.
       (ii) Deduce the molecular formula of the hydrocarbon.   **(WJEC)**

25   (a)   Write an equation for the reaction between hydrogen and chlorine.
   (b)   What volume of product would be obtained by complete reaction of 11.2dm^3 of hydrogen, all gaseous volumes being measured at s.t.p.?   **(O)**

26   142 g of chlorine gas and 168 g of krypton gas (Kr) occupy equal volumes under the same conditions of temperature and pressure. What can you deduce about the gas krypton from this information?   **(C)**

27   Ammonia burns in excess oxygen to give nitrogen and water, according to the equation
$$4NH_3 + 3O_2 = 2N_2 + 6H_2O$$

If 40cm^3 of ammonia are burnt, using 50cm^3 of oxygen, what will be the total volume of the gas after combustion, and what will be its composition by volume? (Assume that all volumes are recorded at the same temperature and pressure, and that the water condenses completely to liquid.)   **(O & C)**

28   (a)   Write down the equation for the complete combustion of methane, CH_4.
   (b)   (i)   Calculate the volume of air, assumed to consist of one-fifth oxygen, that would be required for the complete combustion of 20cm^3 of methane. (All volumes measured at 20 °C and atmospheric pressure.)
       (ii) What would be the total volume and nature of the gases remaining after combustion?   **(O)**

29   (a)   (i)   In the following reaction, state whether the ammonia has acted as an oxidizing agent, a reducing agent or a base:

$$2NH_3 + 3CuO \rightarrow 3Cu + 3H_2O + N_2$$

       (ii) Give reasons for your answer.
   (b)   (i)   If 100cm^3 of ammonia were passed repeatedly over hot copper(II) oxide until reaction was complete, what volume of nitrogen would be produced? (All volumes measured at s.t.p.)
       (ii) What would be the volume of the nitrogen at 25 °C under the same pressure?   **(O)**

30   (a)   Describe *one* test which you would use to show that propene, C_3H_6, is an unsaturated hydrocarbon.
   (b)   (i)   Describe how propene could be converted into propane, C_3H_8.
       (ii) Give the equation for this reaction.
   (c)   Calculate the number of molecules in 560cm^3 of propane gas at s.t.p.
   (d)   (i)   Give the equation for the reaction which takes place when propane is burnt in a plentiful supply of air.
       (ii) Calculate the volume of oxygen which would be used to burn completely 500cm^3 of propane gas. (Assume that all volumes are measured at the same temperature and pressure.)   **(O & C)**

31 This question concerns an experiment to determine the formula of a solid
    hydrocarbon Z.

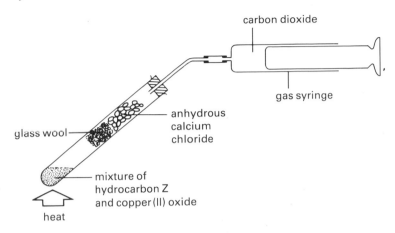

In the apparatus above, 0.016 g of the hydrocarbon Z was *completely* oxidized by
heating with copper(II) oxide. After cooling to room temperature, the volume of
carbon dioxide in the syringe was 30 cm^3.
(a) (i)   Why is it important to use a considerable excess of copper(II) oxide in the
          mixture?
    (ii) What is the purpose of the anhydrous calcium chloride at the top of the
          tube?
(b) (i)   How many moles of carbon dioxide were formed?
    (ii) How many moles of carbon atoms does this contain?
    (iii) What mass of carbon does this represent?
    (iv) What mass of hydrogen was combined with this mass of carbon?
    (v)   How many moles of hydrogen atoms does this represent?
(c) In another such experiment, it was found that the ratio of carbon atoms to
    hydrogen atoms was 0.007 to 0.005
    (i)   What would be the simplest formula of the hydrocarbon?
    (ii) Given that the hydrocarbon has a relative molecular mass of 178, what is
          its molecular formula?
    (iii) Is this hydrocarbon likely to be saturated or unsaturated?   **(L)**

32 In an experiment to determine the percentage by volume of hydrogen and ethane in
    a mixture of the two, 50 cm^3 of the mixture was burned in excess of oxygen and the
    volume of carbon dioxide produced was found to be 40 cm^3, all volumes being
    measured at room temperature and pressure.
    (a) Write an equation for the reaction of ethane with oxygen.
    (b) What volume of ethane would give 40 cm^3 of carbon dioxide?
    (c) Calculate the percentage by volume of ethane in the mixture.   **(WJEC)**

33 20 cm^3 of ethane and 100 cm^3 of oxygen were exploded and the mixture allowed to
    attain the original room temperature and pressure. What will be the volume of each
    of the remaining gases?

    $$2C_2H_6 + 7O_2 = 4CO_2 + 6H_2O$$ **(SUJB)**

34 10 cm^3 of a hydrocarbon C_2H_x required for complete combustion 30 cm^3 of oxygen
    and produced 20 cm^3 of carbon dioxide. What is the value of x?   **(SUJB)**

35 (a) An organic compound X consisting of carbon, hydrogen and oxygen only is found to contain 60.0 per cent by mass of carbon and 13.3 per cent by mass of hydrogen. Find the empirical formula of X.

    (b) A liquid organic compound Y consists of carbon, hydrogen and oxygen only, and has a relative molecular mass of 148. On treatment with metallic sodium, 14.8 g of Y are found to evolve $2.24 \, dm^3$ of hydrogen measured at s.t.p. What can you deduce from this information about the molecule of Y?   (O & C)

5 Moles in solution

5.1 Solubility of solids

Many reagents used in the laboratory are solutions in water. They are usually called *aqueous* solutions. Different substances have different solubilities in water, but for all solids there is a limit to how much will dissolve. A solution is *saturated* when it contains as much of the dissolved solid as possible at a particular temperature.

✱ The solubility of a solid in water, at a given temperature, is the mass of the solute which will just saturate 100 g of water at that temperature.

In one method for finding solubility, the solute is added (with stirring) to water that is kept at the required temperature in a boiling-tube, until no more will dissolve. When the excess solid has settled, a warm pipette is used to transfer some of the saturated solution to a weighed dish. The dish is reweighed with the solution and then the solvent is gently evaporated, leaving the solute. Finally, the dish with solute is weighed.

Example 5a
The solubility of sodium chloride at 40 °C was determined using the method described. The results were:

Mass of dish   = 32.01 g
Mass of dish + solution = 48.40 g
Mass of dish + solute   = 36.40 g

What is the solubility of sodium chloride at 40 °C?

Answer
Mass of solute in the solution = 36.40 g − 32.01 g
                                            = 4.39 g

Mass of water in the solution $= 48.40\,g - 36.40\,g$
$$= 12.00\,g$$
At $40\,°C$, $4.39\,g$ of sodium chloride dissolves in $12.00\,g$ of water
So the solubility of sodium chloride at $40\,°C$
$$= \frac{4.39\,g}{12.00\,g} \times 100$$
$$= 36.58\,g/100\,g \text{ of water}$$

Questions

1 Calculate the solubility of potassium chloride at $40\,°C$, using the following results:

Mass of dish $= 36.22\,g$
Mass of dish + solution $= 49.88\,g$
Mass of dish + solute $= 39.88\,g$

2 Calculate the solubility of sugar at $20\,°C$, using the following results:

Mass of dish $= 35.50\,g$
Mass of dish + solution $= 81.08\,g$
Mass of dish + solute $= 66.08\,g$

3 An investigation of the solubility of ammonium chloride over a range of temperatures gave the results shown in figure 5.1.

	20 °C	40 °C	60 °C	80 °C
Mass of dish (g)	36.30	36.30	36.30	36.30
Mass of dish + solution (g)	52.76	50.88	53.37	52.86
Mass of dish + solute (g)	40.76	40.88	42.37	42.86

Figure 5.1

(a) For each temperature, calculate the solubility of ammonium chloride.
(b) Draw a graph of solubility (vertical axis) against temperature.
(c) What is the effect of temperature on the solubility of ammonium chloride?

In another method of finding solubility, a weighed sample of solute is put into a boiling-tube and a measured volume of water is added. The mixture is heated until the solute disappears. Then the solution is allowed to cool, while being stirred with a thermometer. The temperature at which crystals first appear is recorded. More water is added and the solute is dissolved again. This solution is cooled and crystallized as before.

Example 5b
The results shown in figure 5.2 were obtained using $2.0\,g$ of potassium chlorate, adding $4\,cm^3$ of water each time.

Volume of water (cm³)	Temperature at which crystals appear (°C)
4	92
8	63
12	48
16	35
20	27

Figure 5.2

What is the solubility (a) at 92 °C and (b) at 63 °C?

Answer
(a) At 92 °C, 2.0 g of potassium chlorate dissolves in 4 cm³ (4 g) of water

So the solubility at 92 °C $= \dfrac{2.0 \text{ g}}{4 \text{ g}} \times 100$

$= 50 \text{ g}/100 \text{ g}$ of water

(b) At 63 °C, 2.0 g of potassium chlorate dissolves in 8 cm³ (8 g) of water

So the solubility at 63 °C $= \dfrac{2.0 \text{ g}}{8 \text{ g}} \times 100$

$= 25 \text{ g}/100 \text{ g}$ of water

Questions
4 (a) From the results in example 5b, calculate the solubility of potassium chlorate (i) at 48 °C, (ii) at 35 °C and (iii) at 27 °C.
   (b) Draw a graph of solubility (on the vertical axis) against temperature. Include the solubilities calculated in example 5b.
   (c) What is the effect of temperature on the solubility of potassium chlorate?

5 The solubility of potassium nitrate was investigated using the method described above. 10.0 g of potassium nitrate was used, adding 5 cm³ of water at a time. The results are shown in figure 5.3.

Volume of water (cm³)	Temperature at which crystals appear (°C)
5	88
10	56
15	41
20	32
25	25
30	21

Figure 5.3

(a) Complete a table showing the experimental results and the calculated solubilities.
(b) Draw a graph of solubility (vertical axis) against temperature.

Solubility curves show the variation of the solubility of a substance with temperature. For most solids, solubility increases with temperature, but this does not always happen to the same extent. Figure 5.4 shows the solubility curves for potassium nitrate and sodium chloride. The solubility of potassium nitrate rises steeply with temperature, but the solubility of sodium chloride hardly changes.

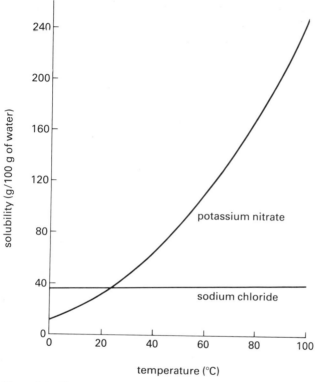

Figure 5.4   How the solubilities of sodium chloride and potassium nitrate vary with temperature

Solubility curves and data can be used to find

* the solubility at other temperatures
* the volume of water that will be needed to dissolve a certain mass of solute at a particular temperature
* the temperature at which a certain mass of solid will dissolve in a fixed volume of water
* the mass of solid that will crystallize when a solution cools or is evaporated.

Example 5c
Using the solubility curve for potassium nitrate in figure 5.4 (obtained from the data in question **5**),
(a) what mass of potassium nitrate can be dissolved in 50 g of water at 20 °C?

(b) what volume of water is needed to dissolve 96 g of potassium nitrate at 20 °C?

(c) what mass of potassium nitrate crystallizes out when a solution containing 50 g of the solid dissolved in 50 cm^3 of water at about 60 °C is cooled to 20 °C?

Answer

(a) From the graph, the solubility of potassium nitrate at 20 °C is about 32 g/100 g of water

So, at 20 °C, the mass which can dissolve in 50 g of water

$$= \frac{32}{100} \times 50\,g$$

$$= 16\,g$$

(b) At 20 °C, the mass of water needed to dissolve 96 g of potassium nitrate

$$= \frac{100}{32} \times 96\,g$$

$$= 300\,g$$

Volume of water
$$= 300\,cm^3$$

(c) At 20 °C, the mass of potassium nitrate dissolved in 50 g of water
$$= 16\,g$$

So the mass crystallized
$$= 50\,g - 16\,g$$
$$= 34\,g$$

Questions

6 Some solubility data for potassium chloride is given in figure 5.5.

Temperature (°C)	10	20	40	60	80	100
Solubility (g/100 g of water)	31.2	34.2	40.0	45.8	51.3	56.3

Figure 5.5

(a) Plot a solubility curve for potassium chloride.

(b) What is the solubility of potassium chloride at 30 °C?

(c) 40 g of potassium chloride is stirred into 100 g of water at 30 °C. Will all the solid dissolve?

(d) At what minimum temperature will 20 g of potassium chloride dissolve in 50 g of water?

(e) What minimum volume of water is needed to dissolve 80 g of potassium chloride at 40 °C?

(f) What mass of potassium chloride can be dissolved in 10 cm^3 of water at 20 °C?

(g) What mass of potassium chloride crystallizes when a saturated solution in 100 cm^3 of water at 60 °C is cooled to 20 °C?

(h) What mass of saturated potassium chloride solution is made by dissolving the solute in 100 g of water at 40 °C?

(i) What mass of potassium chloride crystallizes when 70 g of solution that is saturated at 40 °C is cooled to 20 °C?

7   The solubility of sodium chloride at 20 °C is 36.0 g/100 g of water, and at 100 °C it is 39.8 g/100 g of water.

(a) How much sodium chloride will dissolve in 200 g of water at 20 °C?

(b) If the solution in (a) is heated to 100 °C, how much more sodium chloride will dissolve?

8   Some solubility data for hydrated copper(II) sulphate is given in figure 5.6.

Temperature (°C)	10	20	40	60	80	100
Solubility (g/100 g of water)	17.4	20.7	28.5	40.0	55.0	75.4

Figure 5.6

(a) Plot a solubility curve for copper(II) sulphate.

(b) What mass of copper(II) sulphate is needed to saturate 50 g of water at 70 °C?

(c) What mass of copper(II) sulphate is needed to saturate 50 g of water at 20 °C?

(d) What mass of copper(II) sulphate crystallizes when the saturated solution in (b) is cooled to 20 °C?

(e) What mass of copper(II) sulphate crystallizes when the solution in (b) is evaporated to half its volume, and then cooled to 20 °C?

(f) A 0.5 g seed crystal of copper(II) sulphate is suspended in 100 cm³ of saturated copper(II) sulphate solution at 20 °C. By how much would the temperature of the solution need to rise to cause the seed crystal to dissolve?

9   Figure 5.7 shows the approximate solubilities of potassium chlorate(V) and potassium chloride at 20 °C and at 80 °C, in g/100 g of water.

	20 °C	80 °C
Potassium chlorate(V)	7	27
Potassium chloride	34	51

Figure 5.7

(a) What mass of potassium chlorate(V) will dissolve in 10 cm³ of water (i) at 20 °C and (ii) at 80 °C?

(b) What mass of potassium chloride will dissolve in 10 cm³ of water (i) at 20 °C and (ii) at 80 °C?

(c) A mixture of 2 g of potassium chlorate(V) and 2 g of potassium chloride is stirred into 10 cm³ of water at 80 °C. Will all the mixture dissolve? (Assume that the salts behave independently of each other.)

(d) The solution in (c) is cooled to 20 °C and white crystals appear.

What mass (if any) of (i) potassium chlorate(V) and (ii) potassium chloride is crystallized?

10 Figure 5.8 shows the approximate solubilities of ammonium chloride and potassium chloride at 20 °C and at 60 °C, in g/100 g of water.

	20 °C	60 °C
Ammonium chloride	37	55
Potassium chloride	34	46

Figure 5.8

A mixture of 40 g of ammonium chloride and 40 g of potassium chloride is dissolved in 100 cm³ of water at 60°C, and the solution is cooled to 20 °C. White crystals appear. What percentage of the crystallized solid is ammonium chloride?

5.2 Concentration

In reactions where one or more of the reactants or products is in solution, the *concentration* of the solution(s) may need to be known. Figure 5.9 shows how

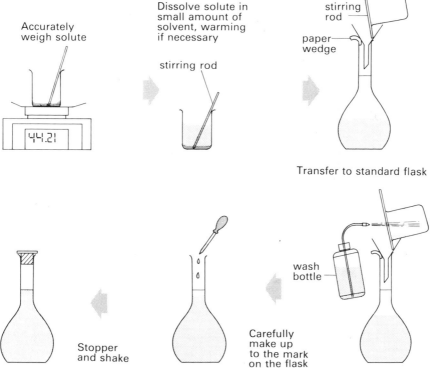

Figure 5.9.   Preparing a solution with a known concentration

a solution with an accurately known concentration can be made by weighing out a sample of solid and dissolving it in water in a graduated flask.

In chemistry, concentrations are usually measured in moles per litre. A solution which contains one mole of solute in 1 dm³ (litre) of solution is labelled $1.0 \, mol/dm^3$ (or $mol \, dm^{-3}$). Chemists sometimes use the term *molar solution* and the abbreviation 1.0 M.

✳ $\begin{array}{c} \text{Amount of solute in a} \\ \text{sample of a solution} \\ \text{(mol)} \end{array} = \begin{array}{c} \text{Volume of} \\ \text{solution} \\ \text{(dm}^3\text{)} \end{array} \times \begin{array}{c} \text{Concentration} \\ \text{(mol/dm}^3\text{)} \end{array}$

Example 5d

What is the concentration of a solution of sodium hydroxide that is made by dissolving 20.0 g of the solid in water and making the volume up to 250 cm³?

Answer

The relative formula mass of sodium hydroxide	= 23 + 16 + 1
	= 40
The molar mass of sodium hydroxide	= 40 g/mol
The amount of sodium hydroxide used	$= \dfrac{20\,g}{40\,g/mol}$
	= 0.5 mol
The volume of the solution	$= \dfrac{250}{1000}\,dm^3$
	= 0.25 dm³
The concentration	$= \dfrac{0.5\ mol}{0.25\ dm^3}$
	= 2.0 mol/dm³

Example 5e

How many moles of sulphuric acid are there in 25 cm³ of a 0.1 M solution?

Answer

The volume of the solution	= 25 cm³
	$= \dfrac{25}{1000}\,dm^3$
The concentration of the solution	= 0.1 mol/dm³
So the number of moles of acid in the sample	$= \dfrac{25}{1000}\,dm^3 \times 0.1\,mol/dm^3$
	= 0.0025 mol

Questions

11   What is the concentration (in mol/dm³) of the following solutions:
(a) 0.1 mol of sodium chloride in 100 cm³ of solution,
(b) 0.5 mol of silver nitrate in 250 cm³ of solution,
(c) 0.002 mol of ammonia in 20 cm³ of solution,
(d) 2.0 mol of zinc sulphate in 4 dm³ of solution?

12   How many moles of the named substance are there in the following:
(a) 25 cm³ of 0.5 M potassium iodide solution,
(b) 100 cm³ of 2 M nitric acid,
(c) 500 cm³ of 0.0001 M copper(II) sulphate solution,
(d) 5 cm³ of 5.0 M potassium carbonate solution?

13   Calculate the mass of solute present in the following samples of
solutions:
(a) 1000 cm³ of 1.0 M hydrochloric acid, HCl,
(b) 10 cm³ of 0.1 M sodium hydroxide, NaOH,
(c) 2 dm³ of 1.0 M sulphuric acid, H_2SO_4,
(d) 25 cm³ of 0.01 M ammonia, NH_3, solution.

14   Imagine that solutions are being made as illustrated in figure 5.9. In
figure 5.10 you are given the names of the solutes and their masses. The
volumes of the graduated flasks used are also stated. For each example,
calculate the concentration of the solution (i) in g/dm³ and (ii) in
mol/dm³.

	Solute	Mass of solute (g)	Volume of flask (cm³)
(a)	Sodium hydroxide, NaOH	4.0	1000
(b)	Silver nitrate, $AgNO_3$	4.25	250
(c)	Potassium iodide, KI	20.75	100
(d)	Potassium manganate(VII), $KMnO_4$	0.79	500

Figure 5.10

5.3   Calculations from equations

Calculations from the equations that involve solutions are carried out in a
similar way to calculations involving the masses of solids or the volumes of
gases.

Example 5f

What mass of solid is precipitated when excess lead(II) nitrate is added to $10\,cm^3$ of $0.5\,mol/dm^3$ potassium iodide solution?

Answer

Step 1 $Pb(NO_3)_2(aq) + 2KI(aq) \rightarrow PbI_2(s) + 2KNO_3(aq)$

Step 2 The lead(II) nitrate is in excess, so the amount of precipitate is determined by the amount of potassium iodide. The equation shows that 2 mol of potassium iodide will produce 1 mol of lead(II) iodide. So 1 mol of potassium iodide will make 0.5 mol of lead(II) iodide

Step 3 The relative formula mass of lead(II) iodide $= 207 + 127$
$$= 334$$

The mass of 1 mol of lead(II) iodide $\quad = 334\,g$

So the mass of lead(II) iodide produced per mole of potassium iodide $= 0.5 \times 334\,g \qquad\qquad = 167\,g$

Step 4 The amount of potassium iodide in this example

$$= \frac{10}{1000}\,dm^3 \times 0.5\,mol/dm^3$$

$$= 0.005\,mol$$

Therefore the mass of precipitate formed

$$= 0.005\,mol \times 167\,g/mol$$
$$= 0.835\,g$$

Example 5g

What volume of gas is given off when excess calcium carbonate is added to $50\,cm^3$ of $2.0\,M$ hydrochloric acid, if the gas volume is measured at s.t.p.?

Answer

Step 1 $CaCO_3(s) + 2HCl(aq) \rightarrow CaCl_2(aq) + CO_2(g) + H_2O(l)$

Step 2 The equation shows that 2 mol of hydrochloric acid produces 1 mol of carbon dioxide, so 1 mol of hydrochloric acid produces 0.5 mol of carbon dioxide

Step 3 The volume of 1 mol of carbon dioxide at s.t.p. is $22\,400\,cm^3$
Thus the reaction will form

$$\frac{22\,400}{2}\,cm^3 = 11\,200\,cm^3 \text{ of gas per mole of hydrochloric acid}$$

Step 4 In $50\,cm^3$ of $2.0\,M$ hydrochloric acid there is

$$\frac{50}{1000}\,dm^3 \times 2\,mol/dm^3 = 0.1\,mol$$

So the volume of gas formed from the amount of acid given in the question $= 0.1\,mol \times 11\,200\,cm^3/mol$
$$= 1120\,cm^3$$

Questions

15 Calculate the mass of silver chloride, AgCl, precipitated when an excess of sodium chloride, NaCl, solution is added to 20 cm^3 of 0.1 M aqueous silver nitrate, $AgNO_3$.

16 Calculate the volume of gas evolved when excess zinc is added to 50 cm^3 of 2.0 mol/dm^3 sulphuric acid, H_2SO_4, if the volume is measured at s.t.p.

17 Calculate the mass of copper(II) oxide, CuO, which will dissolve in 25 cm^3 of 1.0 M sulphuric acid, and calculate the mass of anhydrous copper(II) sulphate, $CuSO_4$, which can be obtained by evaporating the solution to dryness.

18 Calculate the mass of calcium hydroxide required to neutralize 250 cm^3 of 0.5 mol/dm^3 hydrochloric acid.

19 Excess calcium carbonate is added to 25 cm^3 of 4.0 M hydrochloric acid. Calculate (a) the mass of calcium carbonate which reacts and (b) the volume of carbon dioxide formed at s.t.p.

20 Calculate the volume of 1.0 mol/dm^3 hydrochloric acid which can be neutralized by a milk of magnesia tablet containing 0.29 g of magnesium hydroxide.

5.4 Titrations

Titrations are used to investigate reactions and to analyse solutions. A titration involves two solutions. A measured volume of one solution is placed in a flask. The second solution is then added bit by bit from a burette until the reaction is complete.

Investigating reactions

In these experiments, the concentrations of both solutions are known and the aim is to determine the equation for the reaction.

Example 5h

25 cm^3 of 0.1 M sodium carbonate, Na_2CO_3, solution was placed in a flask. A few drops of methyl orange indicator were added and then 0.5 M hydrochloric acid was added from a burette until the indicator changed colour. 10 cm^3 of the acid solution was required. Deduce the left-hand side of the equation for the reaction.

Answer

The amount of sodium carbonate in the flask $= \dfrac{25}{1000} \text{ dm}^3 \times 0.1 \text{ mol/dm}^3$

$$= 0.0025 \text{ mol}$$

The amount of hydrochloric acid added $\quad = \dfrac{10}{1000} \text{ dm}^3 \times 0.5 \text{ mol/dm}^3$

$$= 0.0050 \text{ mol}$$

Thus the ratio is $0.005\,\text{mol}$ $HCl:0.0025\,\text{mol}$ Na_2CO_3
or $2\,\text{mol}$ $HCl:1\,\text{mol}$ Na_2CO_3
The left-hand side of the equation is:

   $2HCl(aq) + Na_2CO_3(aq) \rightarrow$

Questions
21   $10\,\text{cm}^3$ of 0.1 M sodium iodide, NaI, solution reacted exactly with $10\,\text{cm}^3$
     of 0.05 M lead(II) nitrate, $Pb(NO_3)_2$, solution.
     (a) Work out the left-hand side of the equation.
     (b) Given that one of the products is a precipitate of lead iodide, write
         an ionic equation for the reaction, leaving out the spectator ions.
22   Two solutions were made up as follows:
        a colourless solution of barium chloride, $BaCl_2$, $104\,\text{g/dm}^3$
        a yellow solution of potassium chromate(VI), K_2CrO_4, $97\,\text{g/dm}^3$
     Various mixtures of the two solutions were made, and the yellow
     precipitate was separated, washed and dried. The mass of the precipitate
     was measured. The results are shown in figure 5.11.

Experiment	Volume of barium chloride solution (cm³)	Volume of potassium chromate(VI) solution (cm³)	Mass of precipitate (g)
1	50	10	1.27
2	50	20	2.53
3	50	30	3.80
4	50	40	5.06
5	50	50	6.33
6	50	60	6.33
7	50	70	6.33

Figure 5.11

   In the first five experiments, the yellow precipitate was separated from
a colourless solution. In the last two experiments, the solution was also
yellow.
(a) What are the concentrations, in mol/dm^3, of the two solutions?
(b) Plot a graph to show the mass of precipitate formed (vertical axis)
    against the volume of potassium chromate(VI) solution added
    (horizontal axis).
(c) Why was there no increase in the mass of precipitate in the
    experiments numbered 6 and 7? Why was the solution still yellow in
    these last two experiments after the precipitate had been removed?
(d) Work out the left-hand side of the equation.
(e) Given that potassium chloride is soluble in water but that barium
    chromate(VI) is insoluble, suggest a complete equation for the
    reaction.

Analysing solutions

In these experiments, the purpose of the titration is to measure the concentration of an unknown solution, knowing the equation for the reaction and using a second solution of known concentration.

Example 5i

Limewater is a saturated solution of calcium hydroxide, $Ca(OH)_2$, in water. $20\,cm^3$ of limewater was neutralized by $25\,cm^3$ of 0.04 M hydrochloric acid. What was the concentration of the limewater?

Answer

The equation for the reaction is:

$$Ca(OH)_2(aq) + 2\,HCl(aq) \rightarrow CaCl_2(aq) + 2H_2O(l)$$

The amount of hydrochloric acid added $= \dfrac{25}{1000}\,dm^3 \times 0.04\,mol/dm^3$

$$= 0.001\,mol$$

The equation shows that 2 mol of hydrochloric acid reacts with 1 mol of calcium hydroxide

Thus the amount of calcium hydroxide in $20\,cm^3$ of limewater

$$= \dfrac{0.001}{2} = 0.0005\,mol$$

Now let the concentration of limewater be $x\,M$ (i.e. $x\,mol/dm^3$)

So $\dfrac{20}{1000}x = 0.0005$

$$0.02x = 0.0005$$

$$x = \dfrac{0.0005}{0.02}$$

$$x = 0.025$$

Therefore the concentration of the limewater was 0.025 M

Questions

23 $10\,cm^3$ of concentrated hydrochloric acid were diluted to $100\,cm^3$ with water. $10\,cm^3$ of the diluted acid was found to react exactly with $20\,cm^3$ of 1.0 M sodium hydroxide solution.
(a) Write an equation for the reaction.
(b) Work out the concentration of the diluted acid solution from the result of the titration.
(c) Work out the concentration of the undiluted acid.

24 $20\,cm^3$ of 0.1 M sulphuric acid was required to react with $50\,cm^3$ of a solution of sodium hydroxide. Follow these steps to work out the concentration of the alkali:
(a) Write a balanced equation for the reaction.

(b) How many moles of sulphuric acid were added?

(c) According to the equation, how many moles of sodium hydroxide must there have been in $50 \, cm^3$ of the alkali?

(d) How many moles of sodium hydroxide were there in $1 \, dm^3$ ($1000 \, cm^3$) of the alkali?

25 $25 \, cm^3$ of nitric acid, HNO_3, was neutralized by $18.0 \, cm^3$ of $0.15 \, M$ potassium hydroxide, KOH. Calculate the concentration of the nitric acid.

5.5   A formula for titration calculations

There is a formula which some people find helpful for solving problems about titrations. But the formula can only be used correctly if it is understood. Think about the following theoretical example, in which substance A reacts with substance B according to the equation:

$$n_A A + n_B B \rightarrow \text{Products}$$

This means that n_A moles of A reacts with n_B moles of B.

Figure 5.12 shows the apparatus that could be used for a titration involving this reaction. The concentration of the solution of A in the flask is $c_A \, mol/dm^3$. The concentration of the solution of B in the burette is $c_B \, mol/dm^3$. $V_A \, cm^3$ of the solution of A is measured into the flask. Then the solution of B is added from the burette until an indicator shows that the reaction is complete. At this point, $V_B \, cm^3$ has been added.

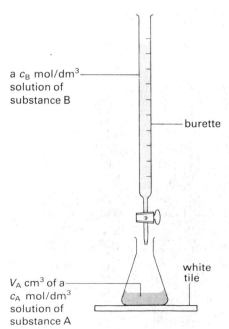

a $c_B \, mol/dm^3$ solution of substance B

burette

white tile

$V_A \, cm^3$ of a $c_A \, mol/dm^3$ solution of substance A

Figure 5.12   Titration apparatus

The amount of A in the flask
$$= \frac{V_A}{1000} \, dm^3 \times c_A \, mol/dm^3$$

$$= \left(\frac{V_A}{1000} \times c_A \right) mol$$

The amount of B added from the burette $= \dfrac{V_B}{1000} \, dm^3 \times c_B \, mol/dm^3$

$$= \left(\frac{V_B}{1000} \times c_B \right) mol$$

But n_A moles of A reacts with n_B moles of B.

Therefore
$$\frac{\left(\dfrac{V_A}{1000} \times c_A \right) mol}{\left(\dfrac{V_B}{1000} \times c_B \right) mol} = \frac{n_A \, mol}{n_B \, mol}$$

This simplifies to

$$* \quad \frac{V_A \times c_A}{V_B \times c_B} = \frac{n_A}{n_B}$$

Investigating reactions
In titrations to investigate reactions, the problem is to determine the ratio n_A/n_B. The concentrations of the two solutions, c_A and c_B, are known and the volumes, V_A and V_B, are measured during the titration. So the desired ratio can be calculated using the formula.

Example 5j
A 41 g sample of phosphonic acid, H_3PO_3, was dissolved in water and the volume of the solution was made up to $1 \, dm^3$. $20 \, cm^3$ of this solution was required to react with $25 \, cm^3$ of a solution of sodium hydroxide. The sodium hydroxide solution contained 32 g of the alkali per litre. What is the equation for the reaction?

Answer
In this question, let the phosphonic acid be compound A and the sodium hydroxide be compound B
The relative formula mass of phosphonic acid $= 3 + 31 + (3 \times 16)$
$$= 82$$

The concentration of the acid,
$$c_A = \frac{41 \, g/dm^3}{82 \, g/mol}$$
$$= 0.5 \, mol/dm^3$$

From the titration readings,
$$V_A = 20 \, cm^3$$

The relative formula mass of sodium hydroxide, NaOH

$$= 23 + 16 + 1$$
$$= 40$$

The concentration of the alkali, $c_B = \dfrac{32\,g/dm^3}{40\,g/mol}$

$$= 0.8\,mol/dm^3$$

From the titration, $V_B = 25\,cm^3$

Substituting in the equation and cancelling the units gives

$$\frac{20 \times 0.5}{25 \times 0.8} = \frac{n_A}{n_B}$$

Thus $\dfrac{n_A}{n_B} = \dfrac{10}{20} = \dfrac{1}{2}$

Therefore the left-hand side of the equation is:

$$H_3PO_3 + 2NaOH \rightarrow$$

An acid is neutralized by a base to form a salt and water. The full equation is therefore:

$$H_3PO_3(aq) + 2NaOH(aq) \rightarrow Na_2HPO_3(aq) + 2H_2O(l)$$

Questions

26 A white precipitate is formed when solutions of potassium chloride, KCl, and lead nitrate, $Pb(NO_3)_2$, are mixed. $40\,cm^3$ of a 0.5 M solution of potassium chloride is required to produce the maximum amount of precipitate from $10.0\,cm^3$ of 1.0 M lead nitrate solution.
(a) Work out the left-hand side of the equation.
(b) All potassium salts and all nitrates are soluble in water. Suggest what the precipitate must be and write a full equation for the reaction.
(c) Calculate the mass of the maximum amount of precipitate formed from $10\,cm^3$ of 1.0 M lead nitrate solution in this reaction.

27 Solution A contains $2.54\,g/dm^3$ of dissolved iodine, I_2. Solution B contains $2.48\,g/dm^3$ of sodium thiosulphate, $Na_2S_2O_3 \cdot 5H_2O$. In a titration it was found that $40\,cm^3$ of solution B was required to react with $20\,cm^3$ of solution A.
(a) Work out the left-hand side of the equation.
(b) Given that one of the products is sodium tetrathionate, $Na_2S_4O_6$, write a complete, balanced equation for the reaction.

28 An acid can be represented by the formula H_xY. One mole of the acid ionizes in solution to produce x moles of hydrogen ions. The relative molecular mass of the acid is 192.
    2.4 g of this acid was dissolved in water and the volume made up to

$250\,cm^3$. $20\,cm^3$ of this solution was neutralized by exactly $25\,cm^3$ of a $0.12\,M$ solution of sodium hydroxide, $NaOH$.

(a) Work out the left-hand side of the equation.

(b) What is the value of x?

29 In a titration $1.0\,M$ nitric acid, HNO_3, was added $1\,cm^3$ at a time to $10\,cm^3$ of $1.0\,M$ potassium carbonate, K_2CO_3, solution. Gas was given off each time the acid was added until a total of $20\,cm^3$ had been added. After that there was no further bubbling.

In a second experiment, $144\,cm^3$ of carbon dioxide gas was given off, at room temperature and pressure, when excess dilute nitric acid was added to $0.828\,g$ of anhydrous potassium carbonate.

(a) Work out the left-hand side of the equation.

(b) Calculate the number of moles of carbon dioxide given off when one mole of potassium carbonate reacts.

(c) Complete the equation for the reaction, given that the other products are a salt and water.

Analysing solutions

In titrations to analyse solutions, the equation for the reaction is already known and so the ratio n_A/n_B is known. The concentration of one of the solutions is known and the volumes, V_A and V_B, are measured during the titration. The problem is to determine the concentration of the other solution.

Example 5k

A solution of a metal carbonate, M_2CO_3, was prepared by dissolving $7.46\,g$ of the anhydrous solid in water to give one litre of solution. $25\,cm^3$ of this solution reacted with $27\,cm^3$ of $0.1\,mol/dm^3$ hydrochloric acid. Calculate the relative formula mass of M_2CO_3 and hence the relative atomic mass of the unknown metal M.

Answer

In this question, let the metal carbonate be compound A and let the hydrochloric acid be compound B

The carbonate ion is CO_3^{2-} so the metal ion in M_2CO_3 must be M^+. Since the chloride ion is Cl^-, the metal chloride is MCl. The equation for the reaction is therefore:

$$M_2CO_3(aq) + 2HCl(aq) \rightarrow 2MCl(aq) + CO_2(g) + H_2O(l)$$

The ratio $\dfrac{n_A}{n_B} = \dfrac{1}{2}$

The volume of the carbonate solution in the titration, $V_A = 25\,cm^3$

The concentration c_A, in mol/dm^3, is not known and has to be calculated

For the hydrochloric acid, $V_B = 27\,cm^3$ and $c_B = 0.1\,mol/dm^3$

Substituting in the titration formula $\dfrac{25 \times c_A}{27 \times 0.1} = \dfrac{1}{2}$

So $$c_A = \dfrac{27 \times 0.1}{2 \times 25}$$

$$= 0.054$$

The concentration of the carbonate solution, $c_A = 0.054 \, mol/dm^3$

Thus the amount of M_2CO_3 in one litre (dm^3) of the solution $= 0.054 \, mol$

This amount of the carbonate has a mass of 7.46 g (as stated in the question)

Therefore the molar mass of the carbonate $\qquad = \dfrac{7.46 \, g}{0.054 \, mol} = 138 \, g/mol$

The relative formula mass of M_2CO_3 $\qquad = 138$

The relative formula mass of the carbonate ion $\qquad = 60$

Therefore the relative atomic mass of the metal M $\qquad = \dfrac{138 - 60}{2}$

$$= 39$$

Questions

30   $25 \, cm^3$ of a solution containing $5.6 \, g/dm^3$ of potassium hydroxide reacted exactly with $15 \, cm^3$ of dilute nitric acid. Calculate the concentration of the nitric acid (a) in mol/dm^3 and (b) in g/dm^3.

$$KOH(aq) + HNO_3(aq) \rightarrow KNO_3(aq) + H_2O(l)$$

31   A 2.65 g sample of anhydrous sodium carbonate was dissolved in water and the solution made up to $250 \, cm^3$. In a titration, $25 \, cm^3$ of this solution required $22.5 \, cm^3$ of hydrochloric acid for neutralization. Calculate the concentration of the acid in mol/dm^3.

32   A 1.575 g sample of ethanedioic acid crystals, $H_2C_2O_4 \cdot nH_2O$, was dissolved in water and made up to $250 \, cm^3$. One mole of the acid is neutralized by two moles of sodium hydroxide. In a titration, $25 \, cm^3$ of this solution of the acid reacted with exactly $15.6 \, cm^3$ of $0.16 \, mol/dm^3$ sodium hydroxide solution.
(a) Calculate the concentration of the acid in mol/dm^3.
(b) Work out the mass of one mole of the acid crystals, and hence calculate the value of n.

Examination questions

1   Calculate the solubility of potassium chloride in water at $15 \, ^\circ C$ from the following data:

Mass of evaporating basin	$= 23.62 \, g$
Mass of basin + saturated solution	$= 53.41 \, g$
Mass of basin + potassium chloride	$= 30.91 \, g$

(SUJB)

2   Use the solubility curves of substances A, B and C shown below to answer the following questions.

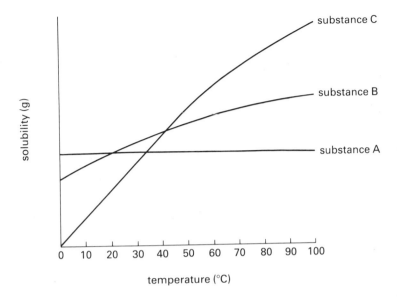

temperature (°C)

(a) Which substance is the most soluble at 100 °C?
(b) Which substance is the most soluble at 10 °C?
(c) Which solubility is affected the most by temperature change?
(d) What would be *seen* if a hot saturated solution of C was cooled rapidly from 90 °C to 30 °C?
(e) What would be *seen* if a crystal of A was put into a saturated solution of A at 20 °C and stirred while the temperature was raised to 80 °C?
(f) What would be *seen* if a crystal of C was added to a saturated solution of C at 20 °C and stirred while the temperature was raised to 80 °C?
(g) A solution containing a mixture of substances A and C was cooled from 90 °C to 20 °C when crystals appeared. Were these crystals A or C?   **(ALSEB)**

3   The table records the masses of different compounds of magnesium, calcium, strontium and barium which dissolve in 100 g of water at 20 °C.

	Chloride, Cl⁻	Sulphate, SO_4^{2-}	Hydroxide, OH⁻
Magnesium, Mg^{2+}	54.2 g	33.0 g	0.0009 g
Calcium, Ca^{2+}	74.5 g	0.21 g	0.156 g
Strontium, Sr^{2+}	53.8 g	0.013 g	0.80 g
Barium, Ba^{2+}	36.0 g	0.0002 g	3.9 g

(a) These four metals are in the same group of the periodic table. In which group are they placed?
(b) What mass of barium chloride would dissolve in 50 g of water at 20 °C to produce a solution which is just saturated?
(c) There are patterns in the solubilities of the sulphates and hydroxides down the group. How do the solubilities change down the group?   **(EAEB)**

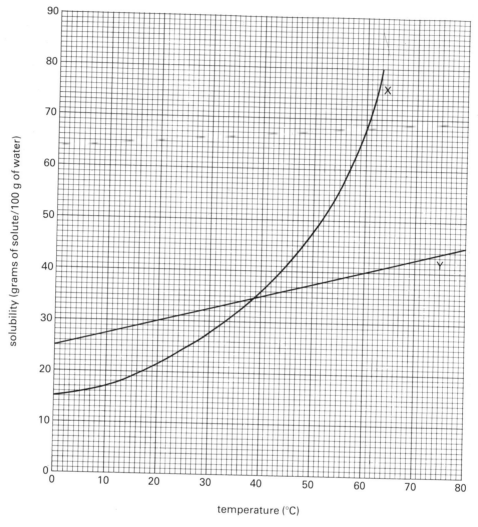

solubility (grams of solute/100 g of water)

temperature (°C)

4 The graph shows the solubility curves of two salts, X and Y. Study the graph and then answer the questions.

(a) What general statement describes the change in solubility of both these salts with a rise in temperature?

(b) How does the solubility of a gas differ from the statement given in your answer to (a)?

(c) At what temperature does X have a solubility of 50 g X/100 g H$_2$O?

(d) At what temperature are the solubilities of X and Y the same?

(e) 100 g of water was saturated with Y at 60 °C and then the solution cooled to 20 °C.

   (i) How many grams of Y would just saturate the 100 g of water at 60 °C?

   (ii) How many grams of Y would just saturate the 100 g of water at 20 °C?

   (iii) How many grams of Y would crystallize on cooling the solution from 60 °C to 20 °C?

(f) Describe how the solubility of a salt may be found experimentally at a particular temperature.

(EAEB)

5 (a) What do you understand by the term *saturated solution*?

(b) The solubilities of potassium chloride and potassium chlorate(V) in water were determined at various temperatures and the results are given in the table below.

Temperature (°C)	0	20	40	60	80	100
Solubility of potassium chloride (g per 100 g of water)	28	34	40	45	51	57
Solubility of potassium chlorate(V) (g per 100 g of water)	3	8	14	24	38	54

From the above data, plot on graph paper the solubility curves of potassium chloride and potassium chlorate(V). (The vertical axis should be used for solubility and the horizontal axis for temperature, and both curves should be plotted between the same axes.)

Use your graphs to answer the following questions:

(i) What is the solubility of potassium chloride at 75 °C?

(ii) What would you observe if a saturated solution of potassium chloride at 75 °C is cooled to 40 °C?

(iii) A saturated solution of potassium chlorate(V) at 80 °C contains 38 g of dissolved solute per 100 g of water. To what temperature must it be cooled in order to reduce the amount of dissolved solute by 19 g?

(iv) Explain what would happen if 50 g of potassium chlorate(V) is added, with stirring, to 100 g of cold water in a beaker and the mixture is heated to boiling point and then allowed to boil vigorously for several minutes.

(v) A mixture of 40 g of potassium chlorate(V) and 40 g of potassium chloride is added to 100 g of water at room temperature and the mixture is heated, with stirring, to 70 °C. The hot mixture is then rapidly filtered, and the filtrate is cooled to 10 °C. Explain what will happen, and calculate the percentage of potassium chlorate(V) in the crystals that are formed at the final temperature. **(AEB)**

6 100 g of water dissolve 32 g of potassium nitrate at 20 °C, and 110 g at 60 °C. If 105 g of a solution of potassium nitrate which is saturated at 60 °C are cooled to 20 °C, what mass of the salt should crystallize out? **(O & C)**

7 (a) (i) Draw a labelled diagram of an apparatus which could be used to show that tap-water contains dissolved gases.

(ii) From such an experiment, 33 cm³ of gases were collected from 1000 cm³ of tap-water. Reaction with white phosphorus caused the volume of these gases to decrease to 22 cm³. Name the two main gases dissolved in tap-water and state their source.

(iii) From the results quoted in (ii), calculate the percentage by volume of each gas.

(iv) Explain why the percentages calculated in (iii) differ from the percentages by volume of these gases present in air.

(b) Describe with full experimental details how you would determine the solubility of potassium nitrate in water at room temperature. State what precautions you would take to obtain an accurate result and what measurements you would make, and show how you would calculate your answer. **(AEB)**

8 The table on the next page gives information about the solubility of the gas hydrogen chloride in water at different temperatures.

(a) Plot a solubility curve for hydrogen chloride in water at different temperatures. Make solubility the vertical axis, and start the scale at 500 g/litre.

Temperature (°C)	0	10	30	40	50	60
Mass (grams dissolving in 1 litre of water)	825	770	675	635	595	560

From your graph determine the number of grams of hydrogen chloride which dissolve in 1 litre of water at 20 °C, and calculate how many moles of hydrogen chloride this represents.

(b) If it is assumed that there is no change in volume when the hydrogen chloride dissolves in water, what is the molarity of the solution at 20 °C?

There is in fact a volume change when the hydrogen chloride dissolves in water. What extra information would you need to enable you to calculate the molarity of the solution *without making the assumption above*?

How would you attempt to determine the molarity of the solution directly, in order to verify your result experimentally?   (L)

9   The solubility of nitrogen monoxide in water at room temperature is one volume to one volume. Given that 1 mole of nitrogen monoxide occupies $24.0 \, dm^3$ at room temperature and pressure, how many moles of the gas will dissolve in $1 \, dm^3$ of water under these conditions? How would an increase in temperature affect the solubility?   (SUJB)

10 A water authority doses a reservoir with iron(III) sulphate at a rate of 10^{-5} moles per dm^3 of water in the reservoir. If the total capacity of the reservoir is 630 million dm^3, calculate

(a)   the total number of moles of iron(III) sulphate required.

(b)   the mass of anhydrous iron(III) sulphate required.   (O)

11 For many solutions their deepness in colour is related to their concentration. Thus dilute solutions of a compound are pale, whereas stronger ones of the same compound are more intense in colour.

This is shown in the following table, which refers to the colour intensity of solutions of an iron compound.

Concentration (grams of Fe/dm^3)	Colour intensity
0.15	21.5
0.20	25.1
0.25	29.4

(The units for colour intensity are not important for this calculation.)

A sample of stainless steel was converted into the same compound as above and dissolved to give a total volume of $250 \, cm^3$. The colour intensity of this solution was found to be 23 when determined as above.

(a)   Plot a graph to show how the colour intensity of these iron solutions varies with concentration.

(b)   Mark clearly on your graph a point corresponding to the solution derived from the stainless steel.

(c)   To what concentration of iron (in g/dm^3) does this point refer?

(d)   If the stainless steel sample used in this experiment actually had a mass of 0.05 g, calculate the percentage of iron present in the alloy.   (O)

12 8 g of magnesium was added to $50 \, cm^3$ of dilute sulphuric acid. After the reaction had ceased the excess metal was recovered, dried and found to have a mass of 1.4 g. What was the molarity of the sulphuric acid?   (SUJB)

13   What mass of solid is precipitated when excess dilute sulphuric acid is added to
      100 cm³ of 2 M barium chloride solution?   (SUJB)

14   What volume of 2 M hydrochloric acid would be required to convert 5.6 dm³ (at
      s.t.p.) of ammonia into ammonium chloride?   (SUJB)

15   Excess magnesium carbonate was added to 25 cm³ of 2 M sulphuric acid. Calculate
      (a) the mass of magnesium carbonate which reacted, and (b) the volume at s.t.p. of
      carbon dioxide evolved.   (SUJB)

16   Excess magnesium carbonate was added to 25 cm³ of 2 M hydrochloric acid. What
      mass of magnesium carbonate would react, and what volume of carbon dioxide,
      measured at s.t.p. would be evolved?   (SUJB)

17   Solution A contains one mole per litre of the chloride of a metal M.
          Solution B contains one mole per litre of silver nitrate ($AgNO_3$).
          In a series of experiments, different volumes of solution B were added to 50 cm³
      portions of solution A and the mass of silver chloride precipitated in each
      experiment was determined. The results of the experiments are shown in the
      following graph.

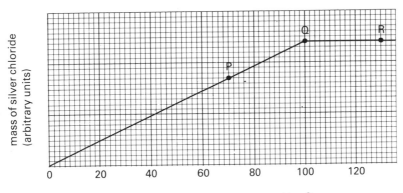

volume of solution B added (cm³)

(a)   Explain why the graph has the shape shown.
(b)   What fraction of one mole of the chloride of M is present in 50 cm³ of solution
      A?
(c)   What fraction of one mole of silver nitrate is present in 100 cm³ of solution B?
(d)   How many moles of silver nitrate are needed to react completely with one mole
      of the chloride of M?
(e)   Write the formula for the chloride of M and the equation for the reaction
      between the chloride and silver nitrate.
(f)   The relative molecular mass of the chloride of M is 111.0. What is the relative
      atomic mass of M?
(g)   Give the formulae of the ions (other than those formed from water) present in
      solution in the liquids corresponding to the points P, Q and R on the graph. (C)

18   In an experiment to determine the molar mass of a monobasic organic acid
      $C_nH_{2n+1}CO_2H$, 0.296 g of the acid was found to react with 20 cm³ of aqueous
      sodium hydroxide of concentration 0.20 mol dm⁻³.
      (a)   Write an equation for the reaction of the acid with aqueous sodium hydroxide.
      (b)   Calculate the number of moles of sodium hydroxide in the solution.
      (c)   From your answers to (a) and (b) calculate the number of moles in 0.296 g of
            the acid.

(d) Calculate the molar mass of the acid.

(e) Find the value of n in the formula $C_nH_{2n+1}CO_2H$.   **(WJEC)**

19 A solution of a monobasic acid X, containing $11.04 \, g/dm^3$ was titrated with a solution of sodium hydroxide containing $8.00 \, g/dm^3$. Using $25.0 \, cm^3$ portions of acid, the burette readings are shown in the table below.

		Rough	Accurate		
Titration number			1	2	3
Final reading (cm^3)		30.7	31.6	31.3	32.1
First reading (cm^3)		0.0	1.6	1.4	2.0
Volume of sodium hydroxide used (cm^3)					

Calculate

(a) the average volume of sodium hydroxide used.

(b) the concentration, in mol/dm^3, of the sodium hydroxide.

(c) the concentration, in mol/dm^3, of X.

(d) the relative molecular mass of X.   **(C)**

20 The formal equation for the neutralization of an aqueous solution of sodium hydroxide by sulphuric acid is:

$$2NaOH + H_2SO_4 = Na_2SO_4 + 2H_2O$$

(a) Write this equation in the simplest possible ionic form.

(b) $25.0 \, cm^3$ of an aqueous solution of sodium hydroxide are found to be neutralized by $22.5 \, cm^3$ of a solution of sulphuric acid containing $9.80 \, g$ of the acid per litre. Calculate (i) the number of moles of NaOH per litre of the sodium hydroxide solution and (ii) the number of grams of NaOH per litre of this solution.   **(O & C)**

21 The equation

$$2HCl + Na_2CO_3 = 2NaCl + H_2O + CO_2$$

represents the reaction between hydrochloric acid and a solution of sodium carbonate as it occurs in a titration using methyl orange as indicator. It was found that $40.0 \, cm^3$ of the acid was neutralized by $25.0 \, cm^3$ of sodium carbonate solution containing 0.40 moles Na_2CO_3 per dm^3.

Using the above data, calculate

(a) the moles of sodium carbonate present in $25.0 \, cm^3$.

(b) the moles of hydrochloric acid present in $40.0 \, cm^3$.

(c) the moles of hydrochloric acid present in $1.00 \, dm^3$.

(d) the concentration of the acid in g HCl per dm^3.

(e) If, instead of hydrochloric acid, a solution of sulphuric acid of the same molar concentration had been used, what volume of this acid would have been required to neutralize $25.0 \, cm^3$ of the sodium carbonate solution?   **(O)**

22 In a titration, $20.0 \, cm^3$ of aqueous barium hydroxide, $Ba(OH)_2$, required $50.0 \, cm^3$ of a solution containing $5.84 \, g/dm^3$ of hydrogen chloride for neutralization. Calculate the concentration of the barium hydroxide solution (a) in mol/dm^3 and (b) in g/dm^3.   **(C)**

23 The bromide of a metal M, of formula MBr_2, reacts quantitatively with silver nitrate in accordance with the following equation:

$$2AgNO_3(aq) + MBr_2(aq) \rightarrow 2AgBr(s) + M(NO_3)_2(aq)$$

In an experiment to determine the relative atomic mass of M, 4.6 g of the bromide were found to react with exactly $100 \, cm^3$ of a solution of silver nitrate of concentration $0.5 \, mol \, l^{-1}$ $(mol \, dm^{-3})$.

(a) Calculate the number of moles of silver nitrate in $100 \, cm^3$ of the solution.

(b) From the answer to (a) and the given equation, calculate the number of moles of the bromide contained in 4.6 g of MBr_2.

(c) From the answer to (b), calculate the molar mass of MBr_2.

(d) Deduce the relative atomic mass of M. **(WJEC)**

24 An alloy of sodium and lead is often used for drying organic solvents because it is safer to use than sodium itself. It reacts with water to give sodium hydroxide solution:

$$Na/Pb(s) + H_2O(l) = NaOH(aq) + \tfrac{1}{2}H_2(g) + Pb(s)$$

In an experiment 2.0 g of alloy was added to about $50 \, cm^3$ of water. When fizzing stopped the sodium hydroxide formed was titrated with 1.0 M hydrochloric acid (solution containing 1.0 mole per dm^3), using a suitable indicator. $8.9 \, cm^3$ of the acid were required. Calculate

(a) the number of moles of HCl used.

(b) the number of moles of NaOH present.

(c) the number of moles of sodium in 2.0 g of the alloy.

(d) the mass of sodium present in 2.0 g of alloy and hence, its percentage composition by mass. **(O)**

6 Moles of electrons

6.1 Electrode reactions

When electrolytes conduct an electric current, chemical changes take place at the electrodes. The electrode reactions depend on the nature of the electrolyte and the material of the electrode, but

✳ at the cathode, the electrode reaction involves *gain* of electrons
✳ at the anode, the electrode reaction involves *loss* of electrons.

Electrolytes include molten salts, and aqueous solutions of salts, acids and alkalis. They contain ions.

Cathode reactions
In many cases, metal ions become metal atoms.

Example 6a
What is the cathode reaction when molten lead(II) bromide is electrolysed using carbon electrodes? How many moles of electrons are gained by lead ions to give one mole of lead?

Answer

Electrolyte	**Electrodes**	**Cathode reaction**
Molten lead(II) bromide, $PbBr_2(l)$	Carbon	$Pb^{2+} + 2e^- \rightarrow Pb$

One mole of lead ions gains two moles of electrons to give one mole of lead atoms

---

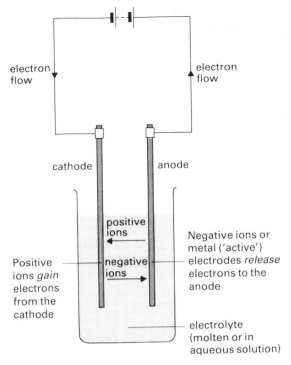

Figure 6.1   Flows of ions in an electrolysis cell

Questions

1 Write equations to show the formation of atoms from the following ions:
   (a) Na^+,   (b) Cu^{2+},   (c) Ni^{2+},   (d) Fe^{3+}.

2 How many moles of electrons are gained by one mole of each of the
   following ions to give one mole of atoms: (a) Na^+,   (b) Cu^{2+},
   (c) Ni^{2+},   (d) Fe^{3+}?

3 How many moles of electrons are needed to give
   (a) 0.2 mol of sodium from sodium ions?
   (b) 0.25 mol of copper from copper(II) ions?
   (c) 0.5 mol of nickel from nickel(II) ions?
   (d) 0.1 mol of iron from iron(III) ions?

4 How many moles of atoms are formed from the following ions by the
   gain of 0.2 mol of electrons: (a) Na^+,   (b) Cu^{2+},   (c) Ni^{2+},
   (d) Fe^{3+}?

5 What mass of the following elements is formed from the ions by the gain
   of 0.2 mol of electrons at the cathode: (a) sodium,   (b) copper,
   (c) nickel,   (d) iron?

During the electrolysis of some aqueous solutions, hydrogen is released at the cathode.

Example 6b

What is the cathode reaction when sodium chloride solution is electrolysed using carbon electrodes? How many moles of electrons are gained by hydrogen ions to give one mole of hydrogen molecules?

Answer

Electrolyte	Electrodes	Cathode reaction
Sodium chloride solution, NaCl(aq)	Carbon	$2H^+ + 2e^- \rightarrow H_2$

Two moles of hydrogen ions gain two moles of electrons to give one mole of hydrogen molecules.

Questions

6 How many moles of electrons are needed to
   (a) change 1 mol of hydrogen ions into hydrogen molecules?
   (b) change 0.2 mol of hydrogen ions into hydrogen molecules?
   (c) form 0.1 mol of hydrogen molecules from hydrogen ions?
   (d) form 2 mol of hydrogen molecules from hydrogen ions?
7 What volume of hydrogen can be obtained at room temperature and pressure
   (a) from 2 mol of hydrogen ions?
   (b) from 0.1 mol of hydrogen ions?
   (c) when hydrogen ions gain 1 mol of electrons?
   (d) when hydrogen ions gain 0.2 mol of electrons?

Anode reactions

Halide ions (chloride, bromide and iodide) become halogens (chlorine, bromine and iodine) by losing electrons at the anode.

Example 6c

What is the anode reaction when copper(II) chloride solution is electrolysed using carbon electrodes? How many moles of electrons are released to the anode by chloride ions to give one mole of chlorine molecules?

Answer

Electrolyte	Electrodes	Anode reaction
Copper(II) chloride solution, $CuCl_2$(aq)	Carbon	$2Cl^- \rightarrow Cl_2 + 2e^-$

Two moles of chloride ions lose two moles of electrons to give one mole of chlorine molecules

Questions

8   How many moles of electrons are released when
    (a) 1 mol of chloride ions become chlorine molecules?
    (b) 0.2 mol of bromide ions (Br^-) become bromine molecules (Br_2)?
    (c) 0.01 mol of iodide ions (I^-) become iodine molecules (I_2)?
    (d) 0.01 mol of chlorine molecules are formed from chloride ions?

9   How many moles of halogen molecules are released by the loss of
    (a) 0.2 mol of electrons from chloride ions?
    (b) 0.5 mol of electrons from chloride ions?
    (c) 0.1 mol of electrons from bromide ions?

10 What volume of chlorine is released at the anode at room temperature
    and pressure when
    (a) 0.4 mol of chloride ions become chlorine molecules?
    (b) 0.1 mol of chloride ions become chlorine molecules?
    (c) chloride ions release 0.02 mol of electrons?

11 What mass of iodine is obtained at the anode when
    (a) 0.2 mol of iodine ions become iodine molecules?
    (b) 0.002 mol of iodide ions become iodine molecules?
    (c) iodide ions release 0.01 mol of electrons?

During the electrolysis of aqueous solutions that do not contain halide ions, oxygen is obtained at the anode.

Example 6d

What is the anode reaction when sodium nitrate solution is electrolysed using carbon electrodes? How many moles of electrons are released to the anode by hydroxide ions to give one mole of oxygen molecules?

Answer

Electrolyte	Electrode	Anode reaction
Sodium nitrate solution, $NaNO_3$(aq)	Carbon	$4OH^- \rightarrow O_2 + 2H_2O + 4e^-$

Four moles of hydroxide ions lose four moles of electrons to give one mole of oxygen molecules and two moles of water

---

Questions

12 How many moles of
    (a) oxygen molecules are released from 1 mol of hydroxide ions?
    (b) oxygen molecules are released when hydroxide ions lose 0.08 mol of electrons?
    (c) electrons are lost to the anode when 0.2 mol of oxygen are formed?
    (d) electrons are lost to the anode by 0.2 mol of hydroxide ions?

13 What volume of oxygen is released at room temperature and pressure when
   (a) 0.004 mol of electrons are lost by hydroxide ions?
   (b) 0.1 mol of electrons are lost by hydroxide ions?
   (c) 0.2 mol of hydroxide ions lose electrons at the anode?

Some metal anodes take part in the anode reaction.

Example 6e
What is the anode reaction when silver nitrate is electrolysed using silver electrodes? How many moles of electrons are released to the anode as one mole of silver atoms become silver ions?

Answer

Electrolyte	**Electrodes**	**Anode reaction**
Silver nitrate solution, $AgNO_3(aq)$	Silver	$Ag \rightarrow Ag^+ + e^-$
		Anode dissolves

One mole of silver atoms loses one mole of electrons to give one mole of silver ions

Question
14 When copper(II) sulphate is electrolysed using copper electrodes, copper atoms in the anode become copper(II) ions in solution.
   (a) Write an equation for the anode reaction.
   (b) How many moles of copper atoms form ions as 0.001 mol of electrons are lost at the anode?
   (c) What mass of copper dissolves from the anode as 0.001 mol of electrons are given up?

6.2 Amounts of electricity

Amounts of electricity (electric charge) are measured in coulombs (C). Current is the rate of flow of electric charge. The practical unit of current is the ampere (A):

✳ $1 A = 1 C/s$

One coulomb of electric charge is used when a current of one ampere flows for one second.
   Useful equations are:

✳ Amount of electric charge (C) = Current (A) × Time (s)

✳ Current (A) = $\dfrac{\text{Amount of electric charge (C)}}{\text{Time (s)}}$

$$* \text{ Time (s)} = \frac{\text{Amount of electric charge (C)}}{\text{Current (A)}}$$

Example 6f
How much electric charge is used when a current of 0.25 A flows for 2 minutes?

Answer
Time of current flow $= 2 \times 60\,\text{s}$
                         $= 120\,\text{s}$

Amount of electric charge $= 0.25\,\text{C/s} \times 120\,\text{s}$
                         $= 30\,\text{C}$

Questions
15 How much electric charge is used when
    (a) a steady current of 0.2 A flows for 600 s?
    (b) a steady current of 0.5 A flows for 20 minutes?
    (c) a steady current of 1 A flows for one hour?
16 For how long do the following currents flow for 500 C of electric charge
    to be used: (a) 1 A,   (b) 0.4 A,   (c) 0.25 A?

6.3 Moles of electrons—the Faraday constant

An electric current is a flow of electrons. Each electron carries a tiny electric charge. The Faraday constant is the charge carried by one mole of electrons. It has a value of about 96 500 C/mol.
   Useful equations involving the Faraday constant are:

$$* \begin{array}{l} \text{Amount of electric} \\ \text{charge (C)} \end{array} = \begin{array}{l} \text{Amount of} \\ \text{electrons (mol)} \end{array} \times \begin{array}{l} \text{Faraday} \\ \text{constant (C/mol)} \end{array}$$

$$* \text{ Amount of electrons (mol)} = \frac{\text{Amount of electric charge (C)}}{\text{Faraday constant (C/mol)}}$$

Example 6g
How many moles of electrons flow in a circuit when a current of 0.4 A is used for 1930 s?

Answer
Amount of electric charge $= 0.4\,\text{C/s} \times 1930\,\text{s}$
                           $= 772\,\text{C}$

Number of moles of electrons $= \dfrac{772\,\text{C}}{96\,500\,\text{C/mol}}$

                           $= 0.008\,\text{mol}$

Questions

17 How much electric charge is carried by
    (a) 0.2 mol of electrons?
    (b) 0.004 mol of electrons?
    (c) 0.05 mol of electrons?

18 How many moles of electrons are needed to carry the following amounts
    of electric charge: (a) 1930 C, (b) 482.5 C, (c) 96.5 C, (d) 28 950 C?

19 How many moles of electrons are involved when
    (a) a current of 0.1 A flows for 1930 s?
    (b) a current of 0.5 A flows for 965 s?
    (c) a current of 0.2 A flows for 2895 s?
    (d) a current of 0.8 A flows for twenty minutes and six seconds?

6.4   How much reaction takes place during electrolysis?

In any electrolysis calculation we may know

✳ the amount (in moles) or mass of product formed
✳ the charge on the ions involved
✳ the amount of electric charge used.

If any two of these are known, the other can be found.

Example 6h

In the electrolysis of molten lead(II) bromide, what mass of lead is formed
at the cathode when a current of 1.0 A flows for 965 s?

Answer

Amount of electric charge used
$$= 1.0\,C/s \times 965\,s$$
$$= 965\,C$$

Number of moles of electrons transferred $= \dfrac{965\,C}{96\,500\,C/mol}$

$$= 0.01\,mol$$

At the cathode:

$$Pb^{2+} + 2e^- \rightarrow Pb$$

   2 mol of electrons is needed to liberate 1 mol of lead

So   0.01 mol of electrons will liberate 0.005 mol of lead

The relative atomic mass of lead $= 207$

The mass of 0.005 mol of lead $= 0.005\,mol \times 207\,g/mol$
$$= 1.035\,g$$

RULES

TO FIND THE AMOUNT OR MASS OF THE PRODUCT

1/ Calculate the amount of electric charge.

2/ Using the Faraday constant, calculate the number of moles of electrons involved.

3/ Knowing the ion charges, write the electrode equation.

4/ Work out the number of moles of product that can be obtained.

5/ Convert moles to mass or volume.

Figure 6.2

Questions

20 How many moles of atoms of the following elements are formed when 965 C of electric charge is used in a suitable electrolysis cell:
(a) copper (Cu^{2+}),   (b) sodium (Na^+),   (c) aluminium (Al^{3+}),
(d) hydrogen (H^+)? (Ion symbols are shown in brackets.)

21 What mass of the following elements is formed when 1930 C of electric charge is used: (a) lead (Pb^{2+}),   (b) nickel (Ni^{2+}),   (c) silver (Ag^+),
(d) iron (Fe^{3+})? (Ion symbols are shown in brackets.)

22 Molten lead(II) bromide is electrolysed using a current of 0.5 A for 1930 s. Use the following steps to find the mass of lead liberated.
(a) How much electric charge is used?
(b) How many moles of electrons are transferred?
(c) Write an equation to show the cathode reaction.
(d) How many moles of lead are formed during the electrolysis?
(e) What mass of lead is formed?

23 What mass of nickel is formed at the cathode when nickel(II) chloride solution is electrolysed using a current of 0.5 A for 772 s?

24 When copper(II) sulphate solution is electrolysed using copper electrodes, copper is deposited at the cathode while the copper anode dissolves, giving copper(II) ions. If a current of 0.2 A flows through the cell for 2895 s
(a) what mass of copper is formed at the cathode?
(b) what mass of copper is lost from the anode?

25 Aluminium is manufactured by the electrolysis of aluminium oxide in molten cryolite. The electric current used is very high and several hundred cells may be connected in series. If the current is 50 000 A

(a) how much electric charge is used in 24 hours?

(b) approximately how many moles of electrons are transferred in 24 hours? (Give your answer to the nearest 1000.)

(c) Using the approximate answer in (b), what mass of aluminium can be produced in one cell in 24 hours? (The aluminium ion is Al^{3+}.)

(d) If 200 cells are connected in series, what mass of aluminium can be produced in 24 hours? Give your answer in tonnes (1 tonne = 1 000 000 g).

26   Chlorine is released at the anode when sodium chloride solution is electrolysed. Using the following steps, find the volume formed at room temperature and pressure when a current of 1 A flows for 965 s.

(a) How much electric charge is used?

(b) How many moles of electrons are transferred in the cell?

(c) Write an equation to show the formation of chlorine molecules (Cl_2) at the anode.

(d) How many moles of chlorine molecules are formed during the electrolysis?

(e) What volume of chlorine is released?

27   During the electrolysis of dilute sulphuric acid using platinum electrodes, hydrogen is released at the cathode and oxygen at the anode. The cathode reaction is:

$$2H^+ + 2e^- \rightarrow H_2$$

The anode reaction is:

$$4OH^- \rightarrow O_2 + 2H_2O + 4e^-$$

If a current of 0.5 A flows for 1930 s and the gases are collected at room temperature and pressure,

(a) what volume of hydrogen is formed at the cathode?

(b) what volume of oxygen is formed at the anode?

Example 6i

When molten magnesium chloride is electrolysed using a current of 0.8 A for 965 s, 0.096 g of magnesium is formed. What is the charge on the magnesium ion?

Answer

Amount of electric charge used
$$= 0.8\,C/s \times 965\,s$$
$$= 772\,C$$

Number of moles of electrons transferred $= \dfrac{772\,C}{96\,500\,C/mol}$

$$= 0.008\,mol$$

Number of moles of magnesium formed $= \dfrac{0.096\,g}{24\,g/mol}$

$$= 0.004\,mol$$

0.004 mol of magnesium is liberated by 0.008 mol of electrons

so   1 mol of magnesium is liberated by 2 mol of electrons

The charge on the magnesium ion is 2+

$$Mg^{2+} + 2e^- \rightarrow Mg$$

TO FIND THE CHARGE ON THE ION

1/ Calculate the amount of electric charge.

2/ Using the Faraday constant, calculate the number of moles of electrons involved.

3/ Calculate the number of moles of the electrode product formed.

4/ Calculate the number of moles of electrons needed to form one mole of the electrode product.

5/ Write the electrode equation.

Figure 6.3

Questions

28   1930 C of electric charge liberates 0.01 mol of cobalt from cobalt nitrate solution.
  (a) How many moles of electrons are gained by 0.01 mol of cobalt ions?
  (b) How many moles of electrons are needed to form one mole of cobalt?
  (c) What is the charge on the cobalt ion?

29   0.238 g of tin can be liberated at the cathode during the electrolysis of tin chloride when a current of 0.5 A flows for 772 s.
  (a) How much electric charge is used?
  (b) How many moles of electrons are gained by tin ions?
  (c) How many moles of tin are formed?
  (d) How many moles of electrons are needed to form one mole of tin?
  (e) What is the charge on the tin ion?

30   A solution of a chromium salt is electrolysed using a current of 0.25 A flowing for 2316 s. The mass of chromium deposited on the cathode is 0.104 g. Calculate the charge on the chromium ion.

Example 6j

For how long, to the nearest minute, will a current of 0.1 A need to flow through a solution of copper(II) sulphate to give 0.032 g of copper at the cathode?

Answer

Number of moles of copper formed $= \dfrac{0.032\,g}{64\,g/mol}$

$\qquad\qquad\qquad\qquad\qquad = 0.0005\,mol$

At the cathode:

$\quad Cu^{2+} + 2e^- \rightarrow Cu$

2 mol of electrons are needed to liberate 1 mol of copper

Number of moles of electrons needed to liberate 0.0005 mol of copper

$\qquad\qquad\qquad\qquad\qquad = 2 \times 0.0005\,mol$

$\qquad\qquad\qquad\qquad\qquad = 0.001\,mol$

Amount of electric charge $\qquad = 0.001\,mol \times 96\,500\,C/mol$

$\qquad\qquad\qquad\qquad\qquad = 96.5\,C$

Time of current flow $\qquad\qquad = \dfrac{96.5\,C}{0.1\,C/s}$

$\qquad\qquad\qquad\qquad\qquad = 965\,s$

$\qquad\qquad\qquad\qquad\qquad \approx 16\ minutes$

RULES

TO FIND THE CURRENT OR TIME

1/ Work out the number of moles of product formed.

2/ Write the electrode equation.

3/ Work out how many moles of electrons are transferred to give the product.

4/ Work out the electric charge.

5/ Work out the current if the time is known, or the time if the current is known.

Figure 6.4

Questions

31 How much electric charge is needed to liberate
  (a) 0.1 mol of silver from silver ions (Ag^+)?
  (b) 0.2 mol of copper from copper ions (Cu^{2+})?
  (c) 0.01 mol of aluminium from aluminium ions (Al^{3+})?
  (d) 0.1 mol of hydrogen (H_2) from hydrogen ions (H^+)?

32 For how long will a current of 1 A flow to liberate 0.001 mol of each of
  the following elements at a cathode:
  (a) silver from silver ions (Ag^+),
  (b) nickel from nickel ions (Ni^{2+}),
  (c) aluminium from aluminium ions (Al^{3+})?

33 A bead of lead of mass 0.414 g is formed at the cathode during the
  electrolysis of molten lead(II) bromide using a current of 0.5 A.
  (a) How many moles of lead are formed?
  (b) How many moles of electrons are needed to liberate the lead?
  (c) How much electric charge is used?
  (d) For how long must the current flow?

34 Hydrogen is liberated at the cathode when dilute sulphuric acid is
  electrolysed. Calculate approximately the current needed to give 96 cm³
  of hydrogen in 16 minutes at room temperature and pressure, using the
  following steps:
  (a) How many moles of hydrogen are formed?
  (b) Write an equation for the cathode reaction.
  (c) How many moles of electrons are needed to give 96 cm³ of
    hydrogen?
  (d) How much electric charge is used?
  (e) What current is needed if the electrolysis time is 16 minutes?

35 What current is needed to plate a cathode with 0.108 g of silver in
  100 minutes? (The electrolyte is a solution containing silver ions, Ag^+.)

36 How long would it take to dissolve away a copper anode of mass 3.2 g
  during the electrolysis of copper(II) sulphate with copper electrodes,
  using a current of 0.5 A?

37 How long would it take to produce 48 cm³ of chlorine at the anode at
  room temperature and pressure when sodium chloride solution is
  electrolysed, using a current of 0.25 A?

When electrolysis cells are connected in series, the same current flows
through each cell and the same amount of electric charge is used.

Example 6k
Two electrolysis cells are connected in series. Cell A contains copper(II)
sulphate solution with copper electrodes. Cell B contains silver nitrate
solution with a silver anode. A stainless steel basin acts as the cathode.
During electrolysis, 0.216 g of silver is formed at the cathode.
(a) What mass of copper forms at the cathode in cell A?
(b) If the current is 0.4 A, how long does the electrolysis take?

Answer
(a) *In cell B*

Number of moles of silver deposited $= \dfrac{0.216\,g}{108\,g/mol}$

$$= 0.002\,mol$$

At the cathode: $Ag^+ + e^- \rightarrow Ag$

So number of moles of electrons gained by silver ions

$$= 0.002\,mol$$

Amount of electric charge used $= 0.002\,mol \times 96\,500\,C/mol$

$$= 193\,C$$

In cell A

Number of moles of electrons gained by copper ions $= 0.002\,mol$

At the cathode: $Cu^{2+} + 2e^- \rightarrow Cu$

So number of moles of copper liberated $= 0.001\,mol$

Mass of copper formed $= 0.001\,mol \times 64\,g/mol$

$$= 0.064\,g$$

(b) Amount of electric charge $= 193\,C$

Electrolysis time $= \dfrac{193\,C}{0.4\,C/s}$

$$= 482.5\,s$$

Questions

38 Two electrolysis cells are connected in series. Cell A contains nickel(II) sulphate solution and nickel electrodes. Cell B contains silver nitrate solution and silver electrodes. During electrolysis 0.059 g of nickel is deposited at the cathode in cell A.
(a) How many moles of nickel are deposited?
(b) Write an equation for the cathode reaction in cell A.
(c) How many moles of electrons are gained by nickel ions in this experiment?
(d) Write an equation for the cathode reaction in cell B.
(e) How many moles of silver are deposited?
(f) What mass of silver is deposited?

39 An electrolysis cell, A, containing molten lead(II) bromide with graphite electrodes is connected in series with cell B, which contains sodium chloride solution with graphite electrodes. During electrolysis, 0.83 g of lead forms at the cathode in cell A.
(a) How many moles of lead are deposited?
(b) Write an equation for the cathode reaction in cell A.
(c) How many moles of electrons are gained by lead ions in this experiment?

(d) Hydrogen is released at the cathode in cell B. Write an equation for the cathode reaction.

(e) How many moles of hydrogen molecules are released in this experiment?

(f) What volume of hydrogen, at room temperature and pressure, is released?

40 An electrolysis cell, A, containing copper(II) sulphate solution with copper electrodes is connected in series with cell B, which contains chromium(III) chloride solution with chromium electrodes. During electrolysis, 0.32 g of copper is plated onto the cathode in cell A in 20 minutes.

(a) How many moles of copper are deposited at the cathode in cell A?

(b) How many moles of electrons are gained by copper ions in this experiment?

(c) How much electric charge is used?

(d) What current flows through the cells?

(e) Write an equation for the cathode reaction in cell B.

(f) How many moles of chromium are deposited at the cathode in cell B?

(g) What mass of chromium is deposited?

(h) How long must the current flow to produce 0.26 g of chromium in cell B?

41 An electrolysis cell, A, containing silver nitrate solution with silver electrodes is connected in series with cell B, containing a solution of the chloride of metal X, with electrodes of that metal. During electrolysis, 0.216 g of silver is formed in cell A and, in the same time, 0.001 mol of X is deposited at the cathode in cell B.

(a) How many moles of silver are deposited in cell A?

(b) How many moles of electrons are gained by silver ions in this experiment?

(c) How many moles of electrons are gained by one mole of X ions?

(d) What is the charge on the X ion?

42 An electrolysis cell containing copper(II) sulphate solution is connected in series with a cell containing sodium chloride solution. Graphite electrodes are used in both cells. During electrolysis, 0.16 g of copper is deposited at the cathode in the cell containing copper(II) sulphate. What volume of chlorine, at room temperature and pressure, is released at the anode in the cell containing sodium chloride?

Examination questions

1 A pure specimen of molten lead(II) bromide was electrolysed in a suitable apparatus using inert electrodes.

(a) Give the formulae of the ions present in the liquid.

(b) What changes would you expect to see at each of the electrodes?

(c) Write ionic equations to show the changes taking place at each of the electrodes.

(d) In such an experiment, a current of 0.2 amperes was passed through the molten lead(II) bromide for four minutes.
  (i) What quantity of electricity passed?
  (ii) What would be the mass of the product liberated at the negative electrode?
  (iii) Assuming all the product liberated at the positive electrode to be in the form of a gas, what volume would it occupy at room temperature and atmospheric pressure? **(L)**

2 Electric charge equivalent to 1 mole of electrons (1 faraday) is passed through fused sodium chloride, using carbon electrodes.
  (a) Explain what happens at the anode and state the mass (in grams) of the product.
  (b) Explain what happens at the cathode and state the mass (in grams) of the product. **(O & C)**

3 Aluminium is extracted by the electrolysis of a solution of alumina in molten cryolite. Information about the two materials is given in the table below.

Raw material	Formula	Melting point (°C)
Alumina	Al_2O_3	2040
Cryolite	Na_3AlF_6	1000

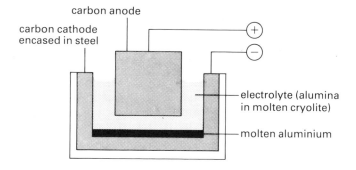

carbon anode

carbon cathode encased in steel

electrolyte (alumina in molten cryolite)

molten aluminium

The alumina can be considered to be ionized:

$$Al_2O_3 \rightarrow 2Al^{3+} + 3O^{2-}$$

(a) Why do you think it is too expensive to extract aluminium by the electrolysis of molten alumina?

(b) Write equations for the production of aluminium metal and oxygen gas (O_2).
  (i) At the anode:
  (ii) At the cathode:

(c) The anode requires frequent replacement. Suggest a reason for this.

(d) A current of 100 amperes produced 2.7 g of aluminium in 5 minutes.
  (i) Calculate the number of coulombs required to deposit 1 mole of aluminium atoms.
  (ii) Use your answer to (d)(i) to show that the charge on the aluminium ion is 3+.

(e) During the electrolysis of aluminium oxide, oxygen is produced. What is the maximum theoretical volume of oxygen gas, measured at room temperature and pressure, which would be produced in the same time as 2.7 g of aluminium?

(f) Aluminium oxide is not soluble in water. How would you demonstrate that aluminium oxide is a base?

(g) Give *three* reasons why aluminium is extremely suitable for cooking utensils such as saucepans.   **(L)**

4 This question concerns the quantity of electricity required to deposit a known mass of copper during the electrolysis of copper(II) sulphate solution, using pure copper electrodes.

Each of four groups of students set up an electric circuit with the electrodes placed in 200 cm³ of 0.5 M copper(II) sulphate solution. The students measured the loss in mass of the anode when a known current was passed for a given time. The students' results are shown below.

Group	Anode loss (g)	Quantity of electricity (coulombs)
1	0.015	45
2	0.040	120
3	0.076	230
4	0.096	290

(a) Draw a labelled diagram of a suitable circuit that the students could have used. The diagram should include all the necessary apparatus.

(b) Plot a graph of the mass of copper lost from the anode against the number of coulombs used.

(c) (i) Use your graph to determine how many coulombs would be required for 0.064 g of copper to be lost from the anode.
      (ii) Calculate how many moles of electrons are needed for one mole of copper atoms to be lost in this experiment.

(d) Write an ionic equation, including state symbols, to show what happens to the copper at the anode during the reaction.

(e) The students in group number 4 used a current of 0.2 amperes. How long, to the nearest minute, was the current flowing in their experiment?   **(L)**

5 This question is about the electrolysis of aqueous copper(II) sulphate solution using copper electrodes. In an experiment it was found that 0.12 g of copper was deposited at the cathode when a current of 0.1 amperes was passed for 1 hour.

(a) Draw a labelled diagram of the electrical circuit you would use in order to measure the quantity of electricity needed to deposit a weighed amount of copper by electrolysis.

(b) State *two* precautions you would take to ensure that the results were as accurate as possible when using the apparatus you have drawn in (a).

(c) Show, by calculation, that the results given above indicate that the charge on the copper ions in this experiment is 2 +.

(d) Write ionic equations, including state symbols, to represent what happens during this electrolysis of aqueous copper sulphate (i) at the cathode and (ii) at the anode.   **(L)**

6 A dilute solution of copper(II) sulphate is electrolysed using platinum electrodes. A current of 0.025 A is maintained until 96.5 C (coulombs) of electricity have passed through the cell.

(a) What changes would you expect to *observe* at the anode, at the cathode and in the solution?

(b) Explain in terms of ions and electrons the changes taking place at the anode and cathode.

(c) For how many seconds must a current of 0.025 A be maintained to pass 96.5 C of electricity?

(d) Calculate the mass of copper deposited during the experiment.

(e) In a similar experiment, it was found that 0.1 g of a metal M having a relative atomic mass of 200 was deposited by 96.5 C. Calculate the mass of metal deposited by one mole of electrons and so deduce the formula of the ions of M present in solution.   (C)

7 Aqueous copper(II) sulphate was electrolysed with copper electrodes. A current of 2 amperes was passed for 8 minutes. During this time the mass of the copper cathode (negative electrode) increased by 0.32 g.

(a) Draw a circuit diagram to show the apparatus you would use to carry out the electrolysis. Label the diagram clearly, and write at the side the names of any other pieces of apparatus which you would use in the experiment.

(b) How many coulombs were passed during the experiment?

(c) Calculate the number of moles of copper atoms deposited during the experiment.

(d) Calculate the number of moles of electrons needed to deposit one mole of copper atoms from aqueous solution.

(e) Write an equation for the reaction at the cathode during electrolysis.   (L)

8 The following results were obtained during the electrolysis of 500 cm³ of aqueous copper(II) sulphate using carbon electrodes.

Total mass of copper deposited on the cathode (g)	Number of coulombs passed
0.33	1 000
1.00	3 000
1.65	5 000
2.30	7 000
3.00	9 000
3.00	11 000
3.00	13 000

(a) Draw a labelled diagram of the apparatus you would use, including the electrical circuit, in order to carry out this electrolysis.

(b) Plot the experimental results on a graph.

(c) Write an ionic equation, including state symbols, to represent the reaction at the cathode during the electrolysis.

(d) Explain the shape of your graph obtained in (b).

(e) Calculate the concentration, in mol/dm³, of the original aqueous copper(II) sulphate.   (C)

9 The pupils in a class were divided into seven groups. Each group was given exactly 200 cm^3 of a green solution of the chloride of a metal M in dilute hydrochloric acid. The groups were asked to electrolyse the solution and measure the quantity of electricity passed.

At the end of the experiment, each group measured the increase in mass of the cathode. The results are tabulated below.

Group	Increase in mass of cathode (g)	Number of coulombs passed	Colour of solution after electrolysis
1	1.50	4 950	Green
2	3.70	12 200	Green
3	7.50	24 750	Pale green
4	9.40	31 000	Very pale green
5	11.71	41 000	Colourless
6	11.69	44 800	Colourless
7	11.70	50 000	Colourless

In all the experiments chlorine gas was evolved from the carbon anode.
(The relative atomic mass of the metal M is 58.5.)

(a) The cathodes were not made of carbon. Suggest a suitable material for the cathodes in these experiments.

(b) What measurements did each group need to make in order to determine the number of coulombs passed?

(c) State, giving brief practical details, how each group should have treated their cathode after removal from the electrolyte but before the final weighing.

(d) Plot the experimental results on a graph. Plot the increase in mass of the cathode on the vertical axis and the number of coulombs passed on the horizontal axis. Join the points with suitable curves or lines.

(e) Explain why groups 6 and 7 did not obtain more of the metal M than group 5.

(f) (i) Calculate the number of moles of electrons required to deposit 1 mole of M on the cathode.

  (ii) What is the formula of the chloride of M? The formula of the chloride ion is Cl^-. Show how you decide on your answer.

(g) Write a balanced equation to show what happens at the *anode* in these experiments. Use the symbol e^- to represent an electron.   **(L)**

10 When a current of 3 amps is passed for 3 minutes 13 seconds through an aqueous solution of one of the sulphates of iron using inert electrodes, 0.112 g of iron is deposited at the cathode (negative electrode).

(a) Calculate the number of moles of electrons required to deposit one mole of iron atoms. Hence determine the size of the charge 'n' for the iron ion Fe^{n+} in the aqueous solution of the sulphate.

(b) Write the formula for this sulphate of iron.

(c) What product(s) would you expect at the anode (positive electrode) during this electrolysis? Write the equation(s) for the reaction(s) taking place at the anode.   **(C)**

11 One faraday of electricity is passed through acidified water. What volume of gas is evolved at each named electrode? (All measurements at s.t.p.)   **(SUJB)**

12 Various molar aqueous solutions were electrolysed using the apparatus below.

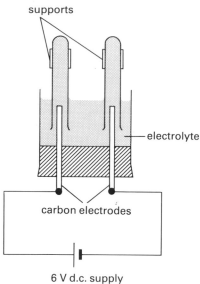

supports

electrolyte

carbon electrodes

6 V d.c. supply

(a) Why were the test-tubes supported above the bung as shown in the diagram?
(b) Complete the table to show the products you would expect when the following solutions are electrolysed.

Solution	Products	
	At cathode	**At anode**
Potassium iodide		
Sodium chloride		
Copper sulphate		

(c) For any *one* of these solutions, give tests that would help to identify the products at the anode and the cathode.
(d) If the solutions are *very* dilute, all three slowly produce the same products at the anode and the cathode. What are these products likely to be?
(e) During the electrolysis of 1 M hydrochloric acid, the reaction at the cathode can be represented by the equation:

$$2H^+(aq) + 2e^- \rightarrow H_2(g)$$

(i) Write an ionic equation, including state symbols, for the production of chlorine molecules from chloride ions at the anode.
(ii) If a current of 0.2 amperes was used for 16 minutes, how many moles of electrons (faradays) would be supplied?
(iii) How many moles of hydrogen molecules would be produced?   **(L)**

13 This question is about the quantitative effect of passing a known current of electricity through the three cells connected in series as shown in the diagram.

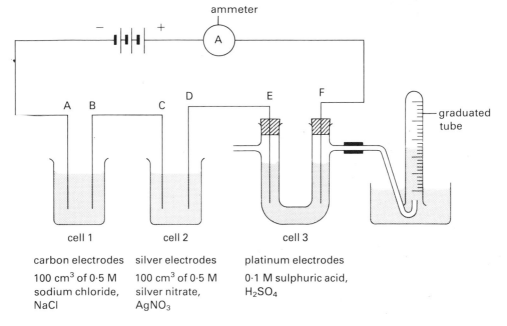

carbon electrodes

100 cm³ of 0·5 M sodium chloride, NaCl

silver electrodes

100 cm³ of 0·5 M silver nitrate, AgNO₃

platinum electrodes

0·1 M sulphuric acid, H₂SO₄

(a) What other piece of apparatus should be placed in the electrical circuit in order to keep the ammeter reading at a steady value?

(b) Which of the electrodes A, B, C, D, E and F are anodes?

(c) Name the substance liberated at the cathode of cell 1.

(d) What is the mass of silver nitrate, $AgNO_3$, present in the solution in cell 2?

The anode in cell 2 was cleaned, dried and weighed both before and after the experiment. It was found to weigh 5.79 g before, and 4.71 g after the experiment. During the experiment a current of 0.2 ampere was passed for 80 minutes.

(e) (i) Describe briefly how the silver anode would be cleaned and dried *after* the experiment.

   (ii) Calculate the number of moles of silver atoms which were removed from the anode during the experiment.

   (iii) Calculate the number of coulombs of electricity used in the experiment.

   (iv) From your answers to (ii) and (iii), calculate the number of moles of electrons required to remove 1 mole of silver atoms from the anode.

(f) The oxygen liberated at electrode F in cell 3 was collected over water as shown. The equation for the discharge of hydroxide ions is:

$$4OH^-(aq) \rightarrow 2H_2O(l) + O_2(g) + 4e^-$$

Write the equation for the discharge of hydrogen ions at the electrode E in the same cell when the same quantity of electricity flows.

(g) Give a brief account of *one* industrial application of electrolysis. **(L)**

7 Kilojoules per mole — energy changes

Every physical or chemical change is accompanied by an energy change.

✱ Processes that absorb energy are described as *endothermic*.
✱ Processes that release energy are described as *exothermic*.

. . . melt about 3 g of ice

. . . warm about 250 cm³ of water by 1°C

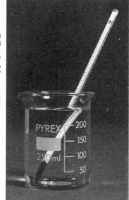

1 kJ of energy can . . .

. . . run an electric kettle for less than ½ second

. . . run a 1 bar electric fire for 1 second

Figure 7.1   Some processes that use 1 kJ of energy

The unit of energy is the joule (J). It is rather a small amount of energy, so energy changes in most processes in chemistry are measured, more conveniently, in kilojoules (kJ). Figure 7.1 gives some idea of what 1 kJ of energy can do.

✱ 1 kJ = 1000 J

The power of an electrical appliance is a measure of the rate at which energy is used. It is measured in watts (W).

✱ 1 W = 1 J/s

A 40 W heater uses 40 J of electrical energy every second and this is converted to heat energy. A 3 kW kettle uses 3 kJ of electrical energy every second.

✱ Power (J/s, i.e. W) = $\dfrac{\text{Energy (J)}}{\text{Time (s)}}$

✱ Energy (J) = Power (J/s, i.e. W) × Time (s)

Example 7a
How much heat energy is supplied by a 100 W heater operating for 2 minutes?

Answer

Heating time
$= 2 \times 60\,\text{s}$
$= 120\,\text{s}$

Energy supplied by heater $= 120\,\text{s} \times 100\,\text{J/s}$
$= 12\,000\,\text{J}$
$= 12\,\text{kJ}$

Questions
1 Convert the following energies in joules (J) to kilojoules (kJ):
  (a) 10 000 J,   (b) 300 J,   (c) 1200 J.
2 Convert the following energies in kilojoules (kJ) to joules (J): (a) 15 kJ,
  (b) 0.25 kJ,   (c) 2.5 kJ.
3 How much heat energy is supplied when
  (a) a 40 W heater is operated for 100 s?
  (b) a 100 W heater is operated for 1 minute?
  (c) a 2 kW kettle is on for 3 minutes?
4 What is the power of
  (a) a heater which supplies 400 J of heat energy in 20 s?
  (b) a heater which supplies 12 kJ of heat energy in 5 minutes?
  (c) an electric kettle which supplies 450 kJ of heat energy in 3 minutes?

7.1 Heating substances

When a substance is heated, the temperature rise depends on the mass and nature of the substance. The specific heat capacity is the amount of energy needed to raise the temperature of 1 g or 1 kg of a substance by 1 °C (1 K). The unit of specific heat capacity is J/(g °C) or kJ/(kg °C).

The specific heat capacity of water is 4.2 J/(g °C).

✱ $\text{Specific heat capacity (J/(g °C))} = \dfrac{\text{Energy supplied (J)}}{\text{Mass (g)} \times \text{Temperature rise (°C))}}$

✱ $\underset{\text{(J)}}{\text{Energy supplied}} = \underset{\text{(g)}}{\text{Mass}} \times \underset{\text{(°C)}}{\text{Temperature rise}} \times \underset{\substack{\text{capacity} \\ \text{(J/(g °C))}}}{\text{Specific heat}}$

Energy is released when a substance cools.

Example 7b

How long will it take to raise the temperature of 100 g of water from 20 °C to 30 °C, using a 40 W heater, assuming that no heat is lost?

Answer

Temperature rise = 30 °C − 20 °C
          = 10 °C

Energy needed = 100 g × 10 °C × 4.2 J/(g °C)
          = 4200 J

Heater supplies energy at the rate of 40 J/s

So heating time $= \dfrac{4200 \text{ J}}{40 \text{ J/s}}$

          = 105 s

Example 7c

A 40 W heater raises the temperature of 100 g of glycerol by 7.5 °C in 45 s. What is the specific heat capacity of glycerol?

Answer

Heater supplies energy at the rate of 40 J/s

So energy used   = 40 J/s × 45 s
          = 1800 J

An energy of 1800 J raises the temperature of 100 g of glycerol by 7.5 °C

Specific heat capacity $= \dfrac{1800 \text{ J}}{100 \text{ g} \times 7.5 \text{ °C}}$

          = 2.4 J/(g °C)

Questions

In these questions, assume that no heat is lost during heating. The specific heat capacity of water is 4.2 J/(g °C).

5 How much energy is needed to raise the temperature of
   (a) 50 g of water by 5 °C?
   (b) 50 g of glycerol by 5 °C? (The specific heat capacity of glycerol is 2.4 J/(g °C).)
6 An energy of 2100 J is transferred to 100 g of water. What is the temperature rise?
7 An energy of 12.6 kJ is transferred to 500 g of water at 22 °C. What is the final temperature of the water?
8 When 1000 g of water cools from 100 °C to 20 °C, how much energy is released from the water?
9 A bunsen burner is used to heat 200 g of water. The temperature rises from 20 °C to 80 °C in 3 minutes.
   (a) How much energy is transferred to the water?
   (b) What is the rate of energy supply by the bunsen burner?
10 A 30 W heater is used to heat 100 g of water. What temperature rise will occur in 70 s?
11 250 g of water is heated using a 200 W heater. How long will it take to raise the temperature of the water from 20 °C to its boiling point?
12 A 40 W heater raises the temperature of 100 g of ethanol by 8 °C in 50 s. What is the specific heat capacity of ethanol?
13 Equal masses of water and ethanol are heated for the same time with the same heater. If the temperature rise of the water is 5 °C, what is the temperature rise of the ethanol if the specific heat capacity ethanol is 2.5 J/(g °C)?

7.2 Physical changes

When a solid reaches its melting point, the temperature stays constant and energy is used to melt the solid. At the boiling point of a liquid, the temperature again remains constant and energy is used to vaporize the liquid. Melting and vaporizing are endothermic processes. Condensing and freezing are exothermic processes. These physical changes are summarized in figure 7.2.

   The amount of energy that is needed to melt a substance at its freezing point or to vaporize a substance at its boiling point depends on the forces of attraction between particles. Substances can be compared by finding the energy needed for one mole of the substances to change.

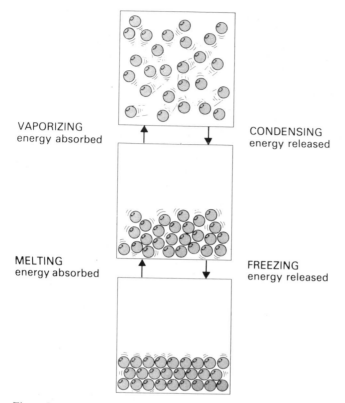

Figure 7.2   The energy changes that accompany physical changes

Example 7d
A 200 W heater vaporizes 9 g of water at its boiling point in 110 s. How
much energy is needed to vaporize one mole of water at its boiling point?

Answer

Energy transferred to water

$$= 200\,\text{J/s} \times 110\,\text{s}$$
$$= 22\,000\,\text{J}$$

Relative molecular mass of water $= 18$

The amount of water vaporized

$$= \frac{9\,\text{g}}{18\,\text{g/mol}}$$
$$= 0.5\,\text{mol}$$

Energy needed to vaporize one mole of water $= \dfrac{22\,000\,\text{J}}{0.5\,\text{mol}}$

$$= 44\,000\,\text{J/mol}$$
$$= 44\,\text{kJ/mol}$$

The data book value is 40.7 kJ/mol

Questions

14 0.5 mol of hexane, C_6H_{14}, is vaporized at its boiling point by an energy of 14 300 J. How much energy is needed to vaporize one mole of hexane at its boiling point?

15 4.6 g of ethanol, C_2H_5OH, is vaporized at its boiling point by an energy of 3900 J.
(a) What is the mass of one mole of ethanol?
(b) How many moles of ethanol are vaporized?
(c) How much energy is needed to vaporize one mole of ethanol?

16 4 g of bromine, Br_2, is vaporized at its boiling point using an energy of 750 J.
(a) What is the mass of one mole of bromine, Br_2?
(b) How many moles of bromine molecules are vaporized?
(c) How much energy is needed to vaporize one mole of bromine molecules at the boiling point?

17 Using the apparatus shown in figure 7.3, 2.1 g of cyclohexane, C_6H_{12}, had condensed when the joulemeter reading was 875 J. (The cyclohexane was boiling steadily before the joulemeter was 'zeroed' and collection started.)
(a) What is the mass of one mole of cyclohexane?
(b) How many moles of cyclohexane were vaporized and condensed?
(c) How much energy would be needed to vaporize one mole of cyclohexane at its boiling point?
(d) Why is the experimental value higher than the data book value of 30 kJ/mol?

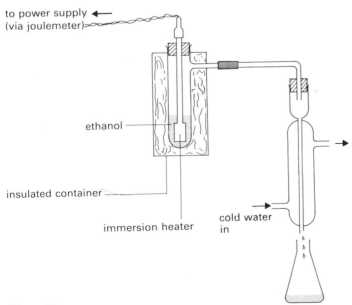

to power supply ◄──
(via joulemeter)

ethanol

insulated container

immersion heater

cold water in

Figure 7.3

18 2.9 g of propanone, CH_3COCH_3, is vaporized at its boiling point by a
40 W heater in 40 s.
(a) How much energy is transferred to the propanone?
(b) What is the mass of one mole of propanone?
(c) How many moles of propanone are vaporized?
(d) How much energy is needed to vaporize one mole of propanone at
its boiling point?

19 The energy needed to vaporize nitrogen, N_2, at its boiling point is
2.8 kJ/mol.
(a) How much energy is needed to vaporize 1 g of nitrogen at its boiling
point?
(b) How long would it take for a 20 W heater to vaporize 1 g of nitrogen
at its boiling point?

20 Two liquids, water and heptane, have nearly the same boiling point. In
separate test-tubes, 1 g of each was placed in a heating bath at 120 °C.
The heptane boiled away in 14 s but the water took 100 s to boil away.
(a) Which liquid needs the most energy to vaporize 1 g?
(b) If about 2300 J of energy is needed to vaporize 1 g of water,
approximately how much energy is needed to vaporize 1 g of
heptane?
(c) The molecular formula of heptane is C_7H_{16}. Approximately how
much energy is needed to vaporize one mole of heptane?

21 The energy needed to vaporize water at its boiling point is 41 kJ/mol.
How much energy is released when steam at 100 °C condenses to give
9 g of water at 100 °C?

22 The energy needed to vaporize ammonia at its boiling point, −33 °C, is
23 kJ/mol. How much energy is released when 340 g of ammonia, NH_3,
is liquefied at its boiling point?

23 An energy of 250 J is used to melt 2.3 g of ethanol at its melting point.
How much energy is needed to melt one mole of ethanol at its melting
point?

24 A 40 W heater is used to melt 27 g of ice at 0 °C to give water at 0 °C in
4 minutes.
(a) How much energy is needed to melt one mole of ice at 0 °C?
(b) How much energy is released when 0.9 kg of water at 0 °C is frozen
to give ice at 0 °C?

25 The melting point of ethanoic acid, CH_3CO_2H, is 17 °C. On a cold day
it is an icy-looking solid. On a warm day it is a colourless liquid. The
energy needed to melt ethanoic acid at its melting point is 12 kJ/mol.
How much energy is needed to melt the 2400 g of glacial ethanoic acid
contained in a stock bottle?

26 A bunsen burner raised the temperature of 100 g of water by 20 °C in
70 s. Without adjusting the flame, it was used to melt 36 g of ice at 0 °C.
(a) How much energy was transferred to the water in 70 s?
(b) What was the rate of energy supply by the bunsen burner?

(c) The energy needed to melt ice at its melting point is 6 kJ/mol.
How much energy is needed to melt 36 g of ice at its melting point?

(d) How long did it take to melt the ice with the bunsen burner?

7.3   Chemical reactions

The heat energy released in exothermic chemical reactions can be found by absorbing this energy in water and measuring the temperature rise. Three ways of doing this are shown in figure 7.4. The container is insulated, if possible, to prevent heat loss from the water or solution.

Endothermic reactions absorb energy from the water or solution, and the temperature falls.

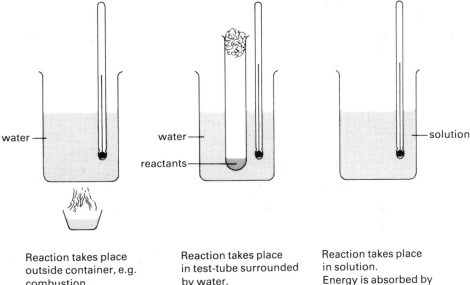

Reaction takes place outside container, e.g. combustion.
Water absorbs the energy

Reaction takes place in test-tube surrounded by water.
Energy is absorbed by or from the water

Reaction takes place in solution.
Energy is absorbed by or from the solution

Figure 7.4   Measuring the heat energy released or absorbed by chemical reactions

Combustion reactions

When fuels burn in air or oxygen, energy is released. All combustion reactions are exothermic. Fuels are compared by finding the energy released by a certain mass of fuel or, more usefully in many cases, by one mole of the fuels.

Example 7e

1.16 g of propanone, CH_3COCH_3, in a crucible burnt away, raising the temperature of 250 g of water in a copper can from 19 °C to 41 °C (see figure 7.5). How much energy is released when one mole of propanone is burnt?

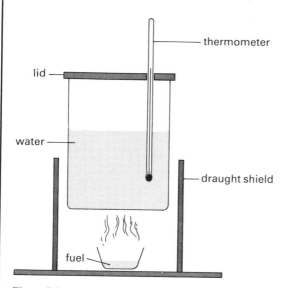

Figure 7.5

Answer

Temperature rise of the water	$= 41\,°C - 19\,°C$
	$= 22\,°C$
Energy absorbed by the water	$= 250\,g \times 22\,°C \times 4.2\,J/(g\,°C)$
	$= 23\,100\,J$
	$= 23.1\,kJ$

The relative molecular mass of propanone

$$= (3 \times 12) + (6 \times 1) + (1 \times 16)$$
$$= 58$$

The amount of propanone burnt $= \dfrac{1.16\,g}{58\,g/mol}$

$$= 0.02\,mol$$

Energy released by burning one mole of propanone

$$= \dfrac{23.1\,kJ}{0.02\,mol}$$

$$= 1155\,kJ/mol$$

The data book value is 1821 kJ/mol

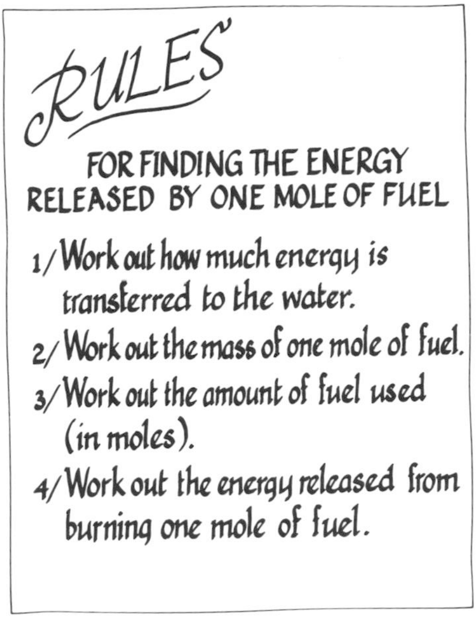

$\mathcal{RULES}$

FOR FINDING THE ENERGY
RELEASED BY ONE MOLE OF FUEL

1/ Work out how much energy is
transferred to the water.

2/ Work out the mass of one mole of fuel.

3/ Work out the amount of fuel used
(in moles).

4/ Work out the energy released from
burning one mole of fuel.

Figure 7.6

Questions

27 0.2 mol of cyclohexane can release an energy of 784 kJ when it is burnt.
How much energy can be obtained from one mole of cyclohexane?

28 3.6 g of butanone, $CH_3CH_2COCH_3$, can release an energy of 122 kJ
when it is burnt.
(a) What is the mass of one mole of butanone?
(b) How many moles of butanone are there in 3.6 g?
(c) How much energy can be released on burning one mole of
butanone?

29 2 g of heptane, C_7H_{16}, can release 97 kJ of energy when it is burnt. How
much energy can be obtained on burning one mole of heptane?

30 Using the apparatus in figure 7.4, 0.92 g of ethanol, C_2H_5OH, raised the
temperature of 250 g of water from 20 °C to 37 °C when it was burnt.
(a) How much energy was transferred to the water?
(b) How many moles of ethanol were burnt?
(c) How much energy would be obtained from one mole of ethanol?

31 The energy available when butane, C_4H_{10}, burns was found by holding
a butane gas cigarette lighter underneath a can containing water. The
lighter was weighed before and after the experiment. 0.29 g of butane
gave a temperature rise of 10 °C in 200 g of water.
(a) How much energy was transferred to the water?
(b) How many moles of butane were used?
(c) How much energy would be obtained from one mole of butane?
(d) Why is this result lower than the data book value of 2877 kJ/mol?

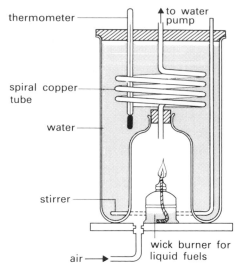

Figure 7.7

32 The apparatus shown in figure 7.7 has a heat capacity of 3.5 kJ/°C. This
means that 3.5 kJ of energy is needed to raise the temperature of the
water and apparatus by 1 °C. When 0.92 g of ethanol was burnt in the
apparatus, the temperature rose by 7 °C.
(a) How much energy was absorbed by the apparatus?
(b) How many moles of ethanol were burnt?
(c) How much energy would be obtained from one mole of ethanol?
(d) Compare this result with the one obtained in question **30**,
    considering the following questions.
    (i) Why are the two results different?
    (ii) Why are both results lower than the data book value of
        1366 kJ/mol?

33 The energy released when methanol, CH_3OH, is burnt is about
22.7 kJ/g. Separate 1 g samples of methanol and of butan-1-ol, C_4H_9OH,
were used in turn to heat a certain volume of water. Using methanol,
the temperature rise was 14 °C. Butan-1-ol gave a temperature rise of
22 °C.
(a) Assuming that the heat losses were the same in both experiments,
    how much energy was released by 1 g of butan-1-ol?
(b) How much energy is available from one mole of butan-1-ol?

34 Methane and hydrogen are gaseous fuels. The energy available from the
combustion of methane is 890 kJ/mol, and that from hydrogen is
286 kJ/mol.
(a) Which fuel gives the most energy per gram?
(b) Which fuel gives the most energy for a given volume?

35   The energies available from the combustion of some alkanes are shown
     in figure 7.8.

Alkane	Molecular formula	Energy released on combustion (kJ/mol)
Methane	CH_4	890
Ethane	C_2H_6	1560
Propane	C_3H_8	2220
Butane	C_4H_{10}	2877
Hexane	C_6H_{14}	4195

Figure 7.8

(a) Draw a bar graph to show the energy released on combustion
    (vertical axis) against the number of carbon atoms in the alkanes.
(b) Predict the energies available from the combustion of one mole of
    (i) pentane and (ii) heptane.

36   The energy available from the combustion of propan-1-ol, C_3H_7OH, is
     about 2020 kJ/mol. A crucible containing 0.6 g of propan-1-ol is used to
     heat 200 g of water in a can.
(a) How much energy is released when 0.6 g of propan-1-ol is burnt?
(b) What temperature rise in the water could result from burning 0.6 g
    of propan-1-ol?
(c) Is the actual temperature rise likely to be lower or higher than the
    calculated one? Why?

37   When coke burns, most of the energy comes from the combustion of
     carbon to give carbon dioxide. The energy available is 393 kJ/mol.
(a) How much energy is released by 1 g of carbon?
(b) How much energy is released from one tonne of coke, assuming that
    it is pure carbon?

38   Natural gas is almost pure methane, CH_4. Most heating in the
     laboratory is done with bunsen burners that use natural gas. The energy
     available from the combustion of natural gas is about 840 kJ/mol.
(a) How much energy is needed to raise the temperature of 500 g of
    water from 20 °C to its boiling point?
(b) How many moles of natural gas are needed to supply the energy?
(c) What volume of natural gas is required at room temperature and
    pressure?
(d) In practice, will the volume of gas used be higher or lower than the
    calculated volume? Why?

Reactions in solution

When magnesium reacts with dilute hydrochloric acid, the solution becomes
warm. The reaction is exothermic. Energy that is released during the reaction is
absorbed by the solution, raising the temperature.

Ammonium nitrate dissolves in water and the solution becomes cold. The process is endothermic. Energy is absorbed from the solution, causing the temperature to fall.

The energy released or absorbed by reactions in solution is found by measuring the temperature rise or fall of the solution in an insulated container such as an expanded polystyrene beaker.

If the solutions are dilute, their specific heat capacities and densities are about the same as those of water. The density of water is about $1 \, g/cm^3$, so the mass of the solution is numerically about the same as its volume.

Example 7f

When 0.48 g of magnesium reacts with $200 \, cm^3$ of $1.0 \, mol/dm^3$ hydrochloric acid, the temperature rises by $10 \, °C$. How much energy is released when the amounts shown in the equation are used (i.e. when one mole of magnesium reacts with hydrochloric acid)?

$$Mg(s) + 2HCl(aq) \rightarrow MgCl_2(aq) + H_2(g)$$

Answer

Number of moles of magnesium used $= \dfrac{0.48 \, g}{24 \, g/mol}$

$$= 0.02 \, mol$$

Number of moles of hydrochloric acid used

$$= \dfrac{200}{1000} \, dm^3 \times 1.0 \, mol/dm^3$$

$$= 0.2 \, mol$$

To react with 0.02 mol of magnesium, only 0.04 mol of hydrochloric acid is needed. So hydrochloric acid is in excess and the energy released depends on the amount of magnesium present

Energy released during the reaction and absorbed by the solution

$$= 200 \, g \times 10 \, °C \times 4.2 \, J/(g \, °C)$$

$$= 8400 \, J$$

Energy released when one mole of magnesium reacts with hydrochloric acid

$$= \dfrac{8400 \, J}{0.02 \, mol}$$

$$= 420\,000 \, J/mol$$

Energy released by the reaction $= 420 \, kJ/mol$

The data book value is $460 \, kJ/mol$

Example 7g

When $50 \, cm^3$ of $2.0 \, mol/dm^3$ hydrochloric acid is added to $50 \, cm^3$ of $2.0 \, mol/dm^3$ sodium hydroxide solution, the temperature rise is $13 \, °C$.

(a) How much energy is released when the amounts (in moles) shown in the equation for the reaction are used?

$$HCl(aq) + NaOH(aq) \rightarrow NaCl(aq) + H_2O(l)$$

(b) What would be the temperature rise if $100\,cm^3$ of $2.0\,mol/dm^3$ hydrochloric acid were added to $100\,cm^3$ of $2.0\,mol/dm^3$ sodium hydroxide solution?

Answer

(a) Number of moles of hydrochloric acid used

$$= \frac{50}{1000}\,dm^3 \times 2.0\,mol/dm^3$$

$$= 0.1\,mol$$

Number of moles of sodium hydroxide used

$$= \frac{50}{1000}\,dm^3 \times 2.0\,mol/dm^3$$

$$= 0.1\,mol$$

Energy released during neutralization and absorbed by the solution

$$= 100\,g \times 13\,°C \times 4.2\,J/(g\,°C)$$
$$= 5460\,J$$

Energy released when one mole of hydrochloric acid reacts with one mole of sodium hydroxide

$$= \frac{5460\,J}{0.1\,mol}$$

$$= 54\,600\,J/mol$$

Energy released by the reaction $= 54.6\,kJ/mol$

The data book value is $57\,kJ/mol$

(b) Number of moles of hydrochloric acid

$$= \frac{100}{1000}\,dm^3 \times 2.0\,mol/dm^3$$

$$= 0.2\,mol$$

Number of moles of sodium hydroxide

$$= \frac{100}{1000}\,dm^3 \times 2.0\,mol/dm^3$$

$$= 0.2\,mol$$

Energy released using $0.2\,mol$ of hydrochloric acid and $0.2\,mol$ of sodium hydroxide

$$= 54\,600\,J/mol \times 0.2\,mol$$
$$= 10\,920$$

Temperature rise
$$= \frac{10\,920\,J}{200\,g \times 4.2\,J/(g\,°C)}$$
$$= 13\,°C$$

Here, twice the amount of substance was used, giving twice the energy output. But the mass of solution was twice as great, so the temperature rise was the same as in the original experiment

RULES

FOR FINDING THE ENERGY
RELEASED OR ABSORBED PER
MOLE OF EQUATION

1/ Write an equation for the reaction.
2/ Work out which reagent is not in excess
   and its amount (in moles).
3/ Work out the energy released or absorbed,
   using the amounts in the experiment.
4/ Work out the energy released or absorbed
   when the amount of reagent (in moles)
   shown in the equation is used.

Figure 7.9

Questions

39 When 0.4 g of calcium reacts with 100 cm³ (excess) dilute hydrochloric acid, a temperature rise of 12 °C occurs. The equation for the reaction is:

$$Ca(s) + 2HCl(aq) \rightarrow CaCl_2(aq) + H_2(g)$$

(a) Is the reaction exothermic or endothermic?
(b) How many moles of calcium were used?
(c) How much energy is released during the reaction and absorbed by the solution?
(d) How much energy is released when one mole of calcium reacts with hydrochloric acid?

40 When 1.3 g of zinc reacts with 100 cm³ of 2.0 mol/dm³ nitric acid, a temperature rise of 6 °C occurs. The equation for the reaction is:

$$Zn(s) + 2HNO_3(aq) \rightarrow Zn(NO_3)_2(aq) + H_2(g)$$

(a) Is the reaction exothermic or endothermic?
(b) How many moles of zinc are used?
(c) How many moles of nitric acid are used?
(d) Which reagent is not in excess?
(e) How much energy is released during the reaction and absorbed by the solution?
(f) How much energy is released when one mole of zinc reacts with nitric acid?

41 When 0.046 g of sodium reacts with 50 cm^3 of water, the temperature of the water rises by about 1.5 °C. How much energy is released when one mole of sodium reacts with water?

42 When 4 g of ammonium nitrate, NH_4NO_3, dissolves in 100 cm^3 of water, the temperature falls by 2.5 °C.
(a) Is the process exothermic or endothermic?
(b) How many moles of ammonium nitrate are dissolved?
(c) How much energy is absorbed from the water when 4 g of ammonium nitrate dissolves in water?
(d) How much energy is absorbed when one mole of ammonium nitrate dissolves in water?

43 When 10.7 g of ammonium chloride, NH_4Cl, dissolves in 200 g of water, the temperature falls by 4 °C.
(a) How many moles of ammonium chloride are dissolved?
(b) How much energy is absorbed from the water when the ammonium chloride dissolves?
(c) How much energy is absorbed when one mole of ammonium chloride dissolves in water?
(d) What fall in temperature is expected when one mole of ammonium chloride dissolves in water, to give 1 dm^3 of a 1.0 mol/dm^3 solution?

44 When 5 g of hydrated copper(II) sulphate, $CuSO_4 \cdot 5H_2O$, dissolves in 100 cm^3 of water, the temperature falls by 0.5 °C. When 3.2 g of anhydrous copper(II) sulphate dissolves in 100 cm^3 of water, the temperature rises by 3 °C.
(a) Is the dissolving of hydrated copper(II) sulphate an exothermic or an endothermic process?
(b) Is the dissolving of anhydrous copper(II) sulphate an exothermic or an endothermic process?
(c) How much energy is released or absorbed when one mole of (i) hydrated and (ii) anhydrous copper(II) sulphate dissolves in water?
(d) If 1.6 g of anhydrous copper(II) sulphate were added to 5 cm^3 of water at 20 °C, what would be the final temperature of the solution?

45 Diluting concentrated sulphuric acid is an exothermic process. When 49 g of concentrated sulphuric acid, H_2SO_4, is added to water to give a final volume of 1000 cm^3, the temperature of the solution rises by 8 °C.
(a) How many moles of sulphuric acid are used?

(b) What is the concentration of the solution that is made?

(c) How much energy is released as the acid is diluted?

(d) How much energy would be released in making a $1 \, mol/dm^3$ solution?

(e) What temperature rise could occur when two moles of sulphuric acid is diluted to $1000 \, cm^3$, giving a $2 \, mol/dm^3$ solution?

46 Zinc displaces copper from copper(II) sulphate solution:

$$Zn(s) + CuSO_4(aq) \rightarrow Cu(s) + ZnSO_4(aq)$$

When 0.5 g of zinc powder is added to $50 \, cm^3$ of $0.2 \, mol/dm^3$ copper(II) sulphate solution, the temperature of the solution rises from 20 °C to 29 °C.

(a) Is the displacement reaction exothermic or endothermic?

(b) How many moles of copper(II) sulphate are used?

(c) How much energy is released during the reaction and absorbed by the solution?

(d) How much energy is released when the amounts shown in the equation are used, i.e. when one mole of copper(II) sulphate is used?

47 Iron displaces copper from copper(II) sulphate solution:

$$Fe(s) + CuSO_4(aq) \rightarrow Cu(s) + FeSO_4(aq)$$

When excess iron filings are added to $25 \, cm^3$ of $0.2 \, mol/dm^3$ copper(II) sulphate solution, a temperature rise of 7 °C occurs.

(a) How many moles of copper(II) sulphate are used?

(b) How much energy is released during the reaction and absorbed by the solution?

(c) How much energy is released when iron reacts with one mole of copper(II) sulphate?

48 Zinc powder displaces silver from silver nitrate solution:

$$Zn(s) + 2AgNO_3(aq) \rightarrow 2Ag(s) + Zn(NO_3)_2(aq)$$

When a slight excess of zinc is added to $50 \, cm^3$ of $0.1 \, mol/dm^3$ silver nitrate solution, a temperature rise of 4 °C occurs.

(a) How many moles of silver nitrate are used?

(b) How much energy is released during the reaction and absorbed by the solution?

(c) How much energy is released when the amounts (in moles) shown in the equation are used?

49 When $25 \, cm^3$ of $0.2 \, mol/dm^3$ silver nitrate solution is added to $25 \, cm^3$ of $0.2 \, mol/dm^3$ sodium chloride solution, a precipitate of silver chloride forms and the temperature of the mixture rises by 1 °C.

$$AgNO_3(aq) + NaCl(aq) \rightarrow AgCl(s) + NaNO_3(aq)$$

(a) How many moles of silver nitrate are used?

(b) How many moles of sodium chloride are used?

(c) How much energy is released in the reaction?

(d) How much energy is released when one mole of silver nitrate reacts with one mole of sodium chloride?

(e) What would be the temperature rise on mixing 50 cm³ of each solution instead of 25 cm³?

50 When 25 cm³ of 1 mol/dm³ calcium chloride solution at 18 °C is added to 25 cm³ of 1 mol/dm³ potassium carbonate solution at 19 °C, a precipitate of calcium carbonate is formed and the temperature of the final mixture is 17.5 °C.

(a) Write an equation for the reaction.

(b) What is the average initial temperature of the solutions?

(c) What is the temperature change during the reaction?

(d) Is the reaction exothermic or endothermic?

(e) How much energy is absorbed from the solution?

(f) How many moles of calcium chloride are used?

(g) How many moles of potassium carbonate are used?

(h) How much energy is absorbed when one mole of calcium chloride reacts with one mole of potassium carbonate in solution?

(i) If sodium carbonate and magnesium chloride solutions are mixed, the same results are obtained. Why?

51 When 25 cm³ of 1.0 mol/dm³ hydrochloric acid is added to 25 cm³ of 1.0 mol/dm³ potassium hydroxide solution, the temperature rise is 6.5 °C.

(a) Write an equation for the reaction.

(b) How many moles of hydrochloric acid are used?

(c) How much energy is released during the neutralization and absorbed by the solution?

(d) How much energy is released when one mole of hydrochloric acid reacts with one mole of potassium hydroxide?

52 When 25 cm³ of 1.0 mol/dm³ sulphuric acid is added to 25 cm³ of 2.0 mol/dm³ sodium hydroxide solution, the temperature rise is 13 °C.

(a) Write an equation for the reaction.

(b) How much energy is released when the amounts (in moles) shown in the equation react?

53 When 25 cm³ of 2.0 mol/dm³ nitric acid is added to 25 cm³ of 1.0 mol/dm³ sodium hydroxide, the temperature rise is 6.5 °C.

(a) Write an equation for the reaction.

(b) One reagent is in excess. Which one?

(c) Which reagent determines the energy released in the reaction?

(d) How much energy is released when one mole of sodium hydroxide is used?

7.4   Energy changes, ΔH and energy level diagrams

All chemical substances have an energy content, H. The energy that is released or absorbed when a physical process or chemical reaction takes place is the change in the energy content. It is the difference in energy content between the reactants and the products. The energy change for the amounts (in moles) shown in the equation is called the *heat of reaction*.

Figure 7.10 shows the energy change in an exothermic reaction. The reactants have an energy content, H_1, and the products have an energy content, H_2. During the reaction, some energy is released by the reactants as they react. So

Figure 7.10   Exothermic reactions

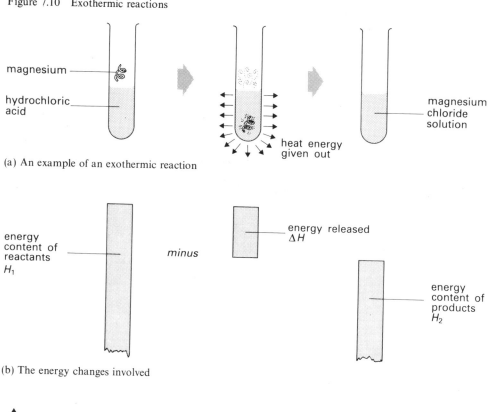

magnesium

hydrochloric acid

heat energy given out

magnesium chloride solution

(a) An example of an exothermic reaction

energy content of reactants H_1

minus

energy released ΔH

energy content of products H_2

(b) The energy changes involved

energy

reactants

ΔH = energy change = heat of reaction

products

(c) Energy level diagram

the energy content of the products, H_2, is lower than that of the reactants, H_1. The heat of reaction (the difference between the energy content of the reactants and the products) can be written as 'ΔH'. 'ΔH' means 'the change in energy content'. It has a sign to show whether the energy content has increased or decreased. For an exothermic reaction, ΔH is negative. The reaction mixture loses energy during the chemical change.

Energy level diagrams like the one in Figure 7.10(c) can be used to show the energy changes.

Figure 7.11 shows the energy change in an endothermic reaction. Energy is taken in by the reaction mixture, so the energy content of the products, H_2, is greater than that of the reactants, H_1. ΔH is positive. In figure 7.11(c), an energy level diagram shows the energy change.

Figure 7.11 Endothermic reactions

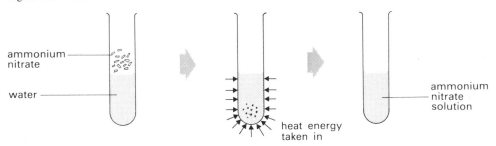

(a) An example of an endothermic reaction

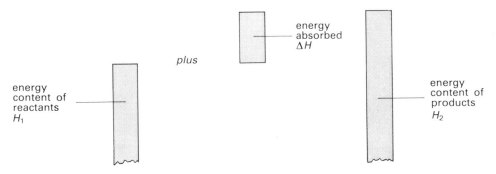

(b) The energy changes involved

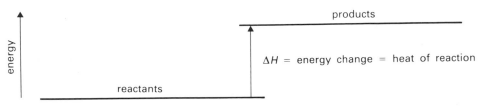

(c) Energy level diagram

All combustion reactions are exothermic. The *heat of combustion* of ethanol is 1366 kJ/mol exothermic. This information can be included in the equation:

$$C_2H_5OH(l) + 3O_2(g) \rightarrow 2CO_2(g) + 3H_2O(l); \quad \Delta H = -1366\,kJ/mol$$

The energy change is for the amounts (in moles) shown in the equation. Figure 7.12 shows the energy level diagram for this reaction.

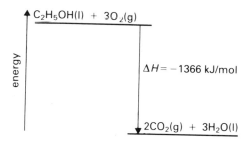

Figure 7.12

All vaporization and melting processes are endothermic. The *heat of vaporization* of water is 41 kJ/mol endothermic:

$$H_2O(l) \rightarrow H_2O(g); \quad \Delta H = +41\,kJ/mol$$

The energy level diagram is shown in figure 7.13.

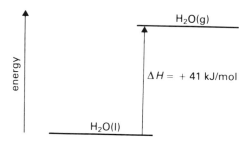

Figure 7.13

* For an exothermic reaction, ΔH is negative.
* For an endothermic reaction, ΔH is positive.

Example 7h
When excess zinc reacts with 0.5 mol of copper(II) ions in solution, 108 kJ of energy is released. What is the heat of reaction, ΔH? Draw an energy level diagram.

Answer
The reaction is exothermic, so ΔH is negative

The equation for the reaction is:

$$Zn(s) + Cu^{2+}(aq) \rightarrow Cu(s) + Zn^{2+}(aq)$$

The energy released when 0.5 mol of Cu^{2+} ions reacts = 108 kJ

$$\text{The energy released when 1 mol of } Cu^{2+} \text{ ions reacts } = \frac{108}{0.5}$$

$$= 216 \text{ kJ}$$

So $\Delta H = -216 \text{ kJ/mol}$

$$Zn(s) + Cu^{2+}(aq) \rightarrow Cu(s) + Zn^{2+}(aq); \qquad \Delta H = -216 \text{ kJ/mol}$$

The energy level diagram is shown in figure 7.14

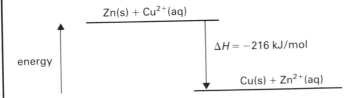

Figure 7.14

Questions

54 When 0.5 mol of water becomes ice, 3.0 kJ of energy is released.
    (a) Write an equation for the process.
    (b) What is the heat of freezing, ΔH?
    (c) Draw an energy level diagram.

55 10.25 kJ of energy is required to change 0.25 mol of water at 100 °C to steam at 100 °C.
    (a) Write an equation for the process.
    (b) What is the heat of vaporization, ΔH?
    (c) Draw an energy level diagram.

56 3.9 kJ of energy is required to change 4.6 g of ethanol, C_2H_5OH, at its boiling point into vapour at the same temperature.
    (a) How many moles of ethanol are vaporized by 3.9 kJ of energy?
    (b) Write an equation for the process.
    (c) What is the heat of vaporization of ethanol, ΔH?
    (d) Draw an energy level diagram.

57 0.86 g of hexane, C_6H_{14}, can release an energy of 41.95 kJ on combustion.
    (a) How many moles of hexane are there in 0.86 g?
    (b) Write an equation to show the combustion of one mole of hexane.
    (c) What is the heat of combustion of hexane, ΔH?
    (d) Draw an energy level diagram.

58   When 0.05 mol of magnesium reacts with hydrochloric acid, an energy of
      23 kJ can be released.
      (a) Write an equation for the reaction.
      (b) What is the heat of reaction, ΔH?
      (c) Draw an energy level diagram.

59   When 80 g of ammonium nitrate, NH_4NO_3, dissolves in 1000 cm³ of
      water in a glass beaker, the temperature of the solution falls by 5°C.
      (a) Write an equation for the process, representing water as 'aq'.
      (b) How much energy is absorbed from the solution?
      (c) What is the heat of solution, ΔH, of ammonium nitrate?
      (d) Draw an energy level diagram.
      (e) Why is the experimental value of ΔH lower than the data book
          value of +25 kJ/mol?

60   When copper displaces silver from a solution containing 0.01 mol of
      silver ions, $Ag^+(aq)$, the energy released is about 700 J.
      (a) Write an equation for the reaction.
      (b) What is the heat of reaction, ΔH, for the amounts (in moles) shown
          in the equation?
      (c) Draw an energy level diagram.

7.5   Why do energy changes occur?

Think about a reaction involving molecules. The reactant molecules become
product molecules. Atoms are rearranged. To do this, the bonds in the reactant
molecules are broken and new bonds in the product molecules are made.

✳ Bond breaking requires energy—it is endothermic.
✳ Bond making releases energy—it is exothermic.

Each type of bond in a molecule has its own 'bond energy'. This amount of
energy is needed to break one mole of the bond, or is released when one mole
of the bond is made. Some 'bond energies' are shown in figure 7.15.

Bond	Bond energy (kJ/mol)
H—H	436
H—O	464
H—Cl	431
H—Br	366
H—I	299
O=O	497
Cl—Cl	242
Br—Br	193
I—I	151
C—H	435
C—Cl	339

Figure 7.15

Example 7i
Using the bond energies given in figure 7.15, calculate the expected energy change for the reaction

$$2H_2(g) + O_2(g) \rightarrow 2H_2O(g)$$

Answer
The equation can be written:

$$2\,H-H + O=O \rightarrow 2 \quad \begin{array}{c} O \\ / \quad \backslash \\ H \quad\quad H \end{array}$$

Figure 7.16 shows the bonds which are broken and made, together with the energy absorbed or released

Bond breaking		Bond making	
Bonds broken	**Energy absorbed (kJ)**	**Bonds made**	**Energy released (kJ)**
2 mol of H—H bonds	872	4 mol of O—H bonds	1856
1 mol of O=O bonds	497		
Energy absorbed =	1369	Energy released =	1856

Figure 7.16

Excess energy released $= 1856\,kJ - 1369\,kJ$
$= 487\,kJ$

Calculated energy change, $\Delta H = -487\,kJ/mol$

Questions
61 Hydrogen burns in chlorine to give hydrogen chloride:

$$H_2(g) + Cl_2(g) \rightarrow 2HCl(g)$$

(a) Draw a structural equation, showing the bonds in the reactant and product molecules.
(b) What bonds are broken in the reaction?
(c) What bonds are made in the reaction?
(d) Draw a table to show the energy absorbed and released as bonds are broken and made. The table should be similar to figure 7.16. Use the bond energies in figure 7.15.
(e) How much energy is released or absorbed during the reaction?

62 Hydrogen burns in bromine to give hydrogen bromide:

$$H_2(g) + Br_2(g) \rightarrow 2HBr(g)$$

Use the bond energies in figure 7.15 to calculate the expected energy change in the reaction.

63 The equation below shows a reaction between methane and chlorine. It
is a substitution reaction. The product is chloromethane.

$$CH_4(g) + Cl_2(g) \rightarrow CH_3Cl(g) + HCl(g)$$

(a) Draw a structural equation to show the bonds in the reactants and
products.
(b) Use the bond energies in figure 7.15 to predict the energy change for
the reaction.
(c) Is the reaction exothermic or endothermic?

Examination questions

In some of these questions you are asked to write on a graph. Do not write in this book
but show your answer on a copy of the graph drawn on your answer paper.

1 An apparatus used for measuring the molar heats of vaporization of liquids of fairly
low boiling point is shown in the diagram. The liquid is heated by a small electric
heater. When the liquid boils, its vapour passes into a water-cooled condenser where
it is condensed to a liquid at room temperature (20 °C). The liquid is collected in a

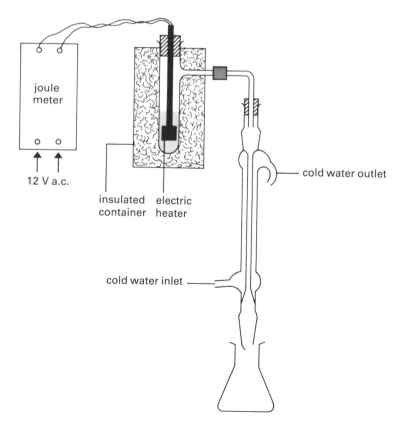

flask and weighed. The energy supplied by the heater to evaporate this mass of liquid is measured by the joulemeter and a value for the molar heat of vaporization can be calculated.

(a) Why would this apparatus be unsuitable for liquids such as ethanal, boiling point 20°C, or pentane, boiling point 36°C?

(b) A second method which is simpler but less accurate is to boil away some liquid from a beaker using a 'steady' bunsen flame, and to compare the mass of liquid lost with the heat input from the bunsen flame. Explain *two* advantages the method described above has over the second simpler method.

(c) An experiment was performed using tetrachloromethane CCl_4, boiling point 77°C, and a heater supplying 40 joules per second. The liquid was heated from 20°C to boiling in 120 seconds, and during 100 seconds of boiling 20 g of tetrachloromethane were collected.

  (i) Calculate the number of kilojoules of energy required to vaporize 1 g of tetrachloromethane.

  (ii) Calculate the number of kilojoules of energy required to vaporize 1 mole of tetrachloromethane molecules.

(d) Draw a graph which shows how temperature varied with time during the experiment described above.   (L)

2 A student performed the following experiment in order to find the heat required to vaporize 1 mole of molecules of trichloromethane, $CHCl_3$ (relative molecular mass may be taken as 120). He placed 250 g of trichloromethane in a suitable apparatus and noted its temperature. He then heated the liquid with a constant bunsen burner flame for 15 minutes, noting the temperature at intervals of time as shown in the table:

Time (minutes)	0	2	4	6	9	12	15
Temperature (°C)	22	37.8	54.5	62	62	62	62

At the end of the experiment, he found that the mass of the remaining trichloromethane was 190 g.

(a) Draw and label a diagram of an apparatus which the student might have used for this experiment. In your diagram show what safety precautions he should have taken to avoid breathing the dangerous trichloromethane vapour.

(b) Plot a graph of the temperature on the vertical axis against time on the horizontal axis.

(c) For how long was the trichloromethane allowed to boil?

(d) How many moles of molecules of trichloromethane were vaporized during the experiment?

(e) The student used these results and from them correctly calculated that the heat required to vaporize 1 mole of molecules of trichloromethane was 40 kJ. The value given in the data book is 29 kJ. If all the student's measurements are accurate, what suggestions can you make to account for this high value?

(f) For most substances which boil below 600°C at 1 atmosphere pressure, there is a relationship between the heat required to vaporize 1 mole of molecules, ΔH(evap), and the boiling point, as shown in the graph on the next page.
      The boiling point of water is 100°C and its ΔH(evap) is 41 kJ mol^{-1}.

  (i) Mark the position of water on the graph.

  (ii) Suggest a reason why water does not behave like trichloromethane and other substances in this respect.   (L)

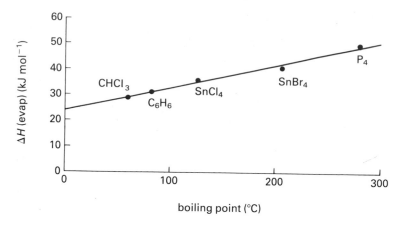

3   The following table of data concerns the changes of state occurring in a number of substances.

Substance	Melting point (°C)	Boiling point (°C)	Molar heat of fusion, ΔH_{fus} (kJ)	Molar heat of vaporization, ΔH_{vap} (kJ)
Methane	−182	−161	0.92	8.2
Hydrogen sulphide	−85	−60	2.50	18.8
Benzene	6	80	9.91	31.0
Sulphuric acid	10.4	326	9.87	50.4
Sodium	97.8	890	2.52	89.0
Lithium	181	1331	2.94	135
Sodium chloride	801	1467	28.6	171

The molar heat of fusion is the energy absorbed by one mole of molecules when it changes from solid to liquid at its melting point.

   The molar heat of vaporization is the energy absorbed by one mole of molecules when it changes from liquid to vapour at its boiling point.
(a)   Which of the above substances would be a liquid at 8 °C?
   The molar heat of vaporization has been plotted opposite against the boiling point for each of the first six substances in the table of data.
(b)   (i) Complete the graph to show the relationship between molar heat of vaporization and boiling point for these six substances.
   (ii) Use the graph to estimate the molar heat of vaporization of xenon, which boils at − 108 °C.
   (iii) What can you deduce from your answer to part (b)(ii) about the bonds between the atoms in liquid xenon? (Note that a xenon molecule consists of only one atom.)
   (iv) Plot the position of sodium chloride on the graph and explain why the values of the molar heats of both fusion and vaporization of sodium chloride are so much greater than the corresponding values for substances such as methane and benzene.

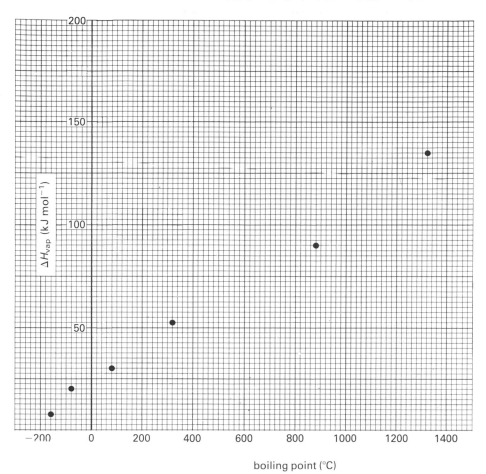

(c) Draw approximately to scale an energy level diagram to represent the following changes of state for benzene, C_6H_6:

$$C_6H_6(s) \rightarrow C_6H_6(l)$$
$$C_6H_6(l) \rightarrow C_6H_6(g)$$

(d) Calculate the quantity of heat in kJ required to melt 234 g of benzene. (The relative molecular mass of benzene, $C_6H_6 = 78$)

An electric heater with an energy output of 1.0 kJ second^{-1} transfers 50% of this energy into benzene in a distillation apparatus. The benzene is brought to the boil and kept boiling for 5 minutes.

(e) (i) Calculate the amount of energy which has entered the boiling benzene during 5 minutes.

(ii) Calculate the mass of benzene which would turn to vapour during this period of boiling.   (L)

4 (a) Propane (C_3H_8) is a gas which is used as a fuel. If it is burnt in a plentiful supply of air,
       (i) name the products of combustion.
       (ii) write an equation for the reaction.
       (iii) state the volume of oxygen which will react completely with 4 litres of propane.
       (iv) assuming that air contains 20% oxygen by volume, what volume of air would be needed for this combustion?
       (v) the heat of combustion is 2200 kJ per mole of propane. Would ΔH for this reaction be given a positive or a negative sign?
       (vi) what mass of propane would need to be burnt in order to obtain 550 000 kJ?
   (b) 0.05 mole of propane occupies 1200 cm^3 at room temperature and pressure.
       (i) What volume would it occupy if the pressure were doubled at constant temperature?
       (ii) If the room temperature were 20 °C, at what temperature would the volume be doubled at constant pressure?   **(L)**

5 (a) Describe, with full practical details, how you would determine the heat of combustion of ethanol. In your answer, you should draw a diagram of a suitable apparatus and state how you would carry out the experiment. Note any observations you would expect to make, any precautions you would take to ensure accuracy and show how you would calculate the result. (Relative molecular mass of ethanol = 46)
   (b) The table below shows the heats of combustion, ΔH, of various alcohols.

Alcohol	Formula	$\Delta H/\text{kJ mol}^{-1}$
Methanol	CH_3OH	−720
Propanol	$CH_3CH_2CH_2OH$	−2010
Butanol	$CH_3(CH_2)_2CH_2OH$	−2670
Pentanol	$CH_3(CH_2)_3CH_2OH$	−3300

       (i) Plot a graph of ΔH, the heat of combustion (y axis), against the number of carbon atoms (x axis).
       (ii) Use your graph to estimate ΔH for ethanol, CH_3CH_2OH, and for hexanol, $CH_3(CH_2)_4CH_2OH$.
       (iii) Give *two* reasons why the experimental results obtained for the heats of combustion are often lower than those quoted in standard tables of data.
                                                                                      (L)

6 Ethanol is sometimes used as a fuel, usually in the form of methylated spirits.
   (a) What is a fuel?
   (b) Give *one* example of the use of ethanol as a fuel.
        The heat of combustion of ethanol, C_2H_5OH, was measured using the simple apparatus shown in the diagram.
        500 g of water were placed in the aluminium container and the temperature of the water was recorded. The small crucible and ethanol were weighed, placed under the container, and the ethanol ignited. The water was kept well stirred. When the temperature of the water had risen about 10 °C the crucible was covered with a fire-resistant mat to put out the flame. When cool the crucible and the remaining ethanol were reweighed.
   (c) Give *one* reason for using an aluminium container for the water instead of a glass beaker.

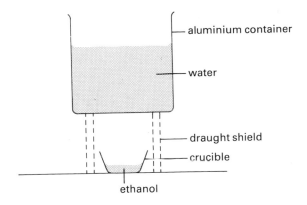

The results of the experiment were:
    Initial mass of crucible and ethanol = 29.96 g
    Final mass of crucible and ethanol = 29.04 g
    Initial temperature of water   = 20 °C
    Final temperature of water   = 30 °C
(d) How many moles of molecules of ethanol were burned?
(e) How many kJ of heat were absorbed by the water?
(f) Calculate the molar heat of combustion of ethanol.
(g) The value of the molar heat of combustion of ethanol given in tables of data is
     greater than the value obtained from this experiment. Explain the reason for
     one possible source of the error in the experiment described.   **(L)**

7 A can containing 100 g of water was heated with a steady bunsen flame. Each
   minute the temperature of the water was noted and the table shows the results for
   the first 6 minutes.

Time (min)	0	1	2	3	4	5	6
Temperature (°C)	20	40	58	79	100	100	100

(a) How long did it take for the water to boil?
(b) At a certain time, someone opened the door, then closed it again. How many
     minutes after the start did this apparently happen? Why have you chosen this
     time?
(c) What was the total rise in temperature?
(d) How many joules were supplied by the bunsen burner up to the time the water
     began to boil? Take

         Joules = Mass of water × Rise in temperature × 4

(e) How many kilojoules is this?
(f) How many kilojoules did the bunsen burner supply in 1 minute on average?
(g) This answer was found to be very low. Describe two ways by which you would
     improve the way in which the experiment was carried out.
(h) If you wanted to use the output per minute of the bunsen burner to find the
     quantity of heat energy needed to turn 1 g of water into steam,
     (i) at what time in minutes would you begin measuring this?
     (ii) state another measurement which would have to be taken other than time
          or temperature.   **(EMREB)**

8 (a) Describe how you would attempt to determine the heat of combustion of ethanol. As part of your answer you should state the measurements you would make and how you would use the results to calculate the heat of combustion in $kJ\,mol^{-1}$.

   The results of experiments to measure the heat evolved when mixtures of ethanol and propan-1-ol were completely burnt in a suitable apparatus are shown below.

Experiment	1	2	3	4	5	6
Number of moles of ethanol in mixture	0.010	0.008	0.006	0.004	0.002	0.000
Number of moles of propan-1-ol in mixture	0.000	0.002	0.004	0.006	0.008	0.010
Total number of moles in mixture	0.010	0.010	0.010	0.010	0.010	0.010
Heat evolved (in kJ)	13.65	14.95	16.25	17.50	18.80	20.10

   (b) Plot a graph of heat evolved (vertical axis) against number of moles of ethanol in each mixture.
   (c) (i) Calculate the heat of combustion of ethanol in $kJ\,mol^{-1}$.
       (ii) Write down the thermochemical equation for the complete combustion of ethanol, indicating the heat evolved.
   (d) 0.010 mole of a mixture of ethanol and propan-1-ol produced 15.60 kJ of heat on combustion.
       (i) From your graph, determine the number of moles of ethanol present in this mixture.
       (ii) Calculate the *mass* of ethanol present in this mixture.   **(L)**

9 This question is about the determination of the heat of solution of sulphuric acid, and the following information may be used in answering the question:

$$H_2SO_4(l) + aq \rightarrow H_2SO_4(aq, 1.0\,M)$$

   In an experiment 0.5 mole of molecules of the pure sulphuric acid was added to sufficient water in a beaker to produce $500\,cm^3$ of a 1.0 M solution. The temperature rose from 20 °C to 37 °C.
   (a) Calculate the mass of sulphuric acid added.
   (b) From these results calculate ΔH for the reaction in the equation given.
   (c) Draw and label a simple energy level diagram for the reaction.
   (d) This experiment is highly simplified and is only expected to produce an approximate value for ΔH. Suggest *one* possible source of error and explain how you could reduce this error by modifying the experiment.   **(L)**

10 This question is about the energy that can be obtained from chemical reactions. The reaction between zinc and copper(II) sulphate solution can be investigated in the apparatus shown below.

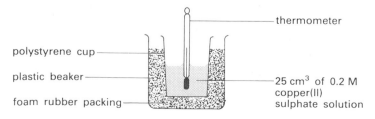

polystyrene cup

plastic beaker

foam rubber packing

thermometer

25 cm³ of 0.2 M copper(II) sulphate solution

An excess of powdered zinc was added to 25 cm³ of a solution containing 0.005 moles of Cu^{2+}(aq) ions. The mixture was stirred with the thermometer and the maximum temperature reached was recorded. The temperature of the mixture was found to rise by 10 °C.

(a)   What would you see when stirring the mixture of copper(II) sulphate solution and zinc?

(b)   Write an equation for this reaction, showing state symbols.

(c)   What is the purpose of the foam rubber packing between the cup and beaker?

(d)   (i)   From the information given, calculate the heat change during the experiment. Show your working and state the units you use.

(ii)   Calculate ΔH for the reaction per mole of Cu^{2+}(aq) ions. Show your working and give the correct sign for ΔH.

(e)   In the above calculation, the heat absorbed by the polystyrene cup is ignored. Explain why this can be done without making the answer too inaccurate.

(f)   Describe, or draw a labelled diagram of, an electrochemical cell in which the oxidation of hydrogen or some other fuel can be used to produce electrical energy.   (L)

11   The apparatus shown in the diagram was used to investigate the reaction between mercury and bromine.

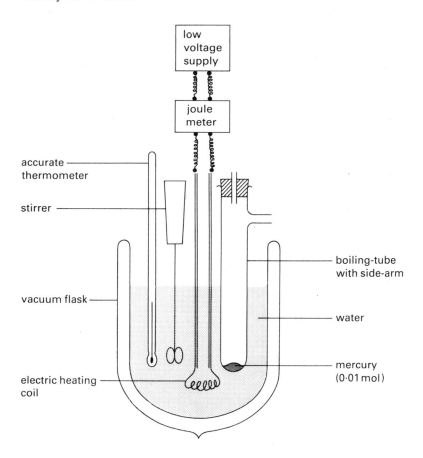

The boiling-tube with side-arm, containing 0.01 mol of mercury, was weighed before the experiment and its mass was 23.08 g. The apparatus was assembled and the temperature of the water measured. The stirrer was switched on and an excess of bromine was added to the mercury. An exothermic reaction took place between the two elements and the maximum temperature rise recorded by the thermometer was noted. The side-arm of the boiling-tube was then connected to a suction pump and the excess of bromine was removed as vapour. A white solid was left in the tube.

Next the low-voltage supply to the heating coil was switched on and it was found that 1.72 kJ were needed to produce the same temperature rise as that produced by the reaction. The side-arm boiling-tube was removed, dried and re-weighed and its mass was 24.68 g.

(a) Give *one* property that is shared by mercury and bromine but which distinguishes them from other elements in the periodic table.

(b) It was found that the excess bromine was removed very slowly by pumping. Suggest one modification of the apparatus that would speed up the removal of the bromine. Write, or draw, your answer clearly on the diagram.

(c) Calculate how many moles of bromine combined with 0.01 mol of mercury (Hg) and give the formula of the white solid formed.

(d) Write a balanced equation for the reaction which took place between mercury atoms and bromine molecules.

(e) Why was the experiment carried out in a vacuum flask?

(f) Why was it important that the water in the apparatus was well stirred during the experiment?

(g) Why did the reading on the thermometer continue to rise for some time after the reaction between the mercury and the bromine had stopped?

(h) Work out the value of ΔH for the reaction in kJ per mole of mercury reacting.

(i) It is possible to estimate ΔH for the reaction in the same apparatus without using a heating coil and joulemeter. The temperature rise produced by the reaction and the mass of water in the flask are both measured and the heat change is calculated given that 4.18 J raises the temperature of 1 g of water by 1 °C.

   (i) If this method had been used, how would the calculated value of ΔH differ from the result obtained using an electric heating coil?

   (ii) Explain why this method is less accurate than the method using a heating coil. **(L)**

12 This question is about the heats of neutralization of acids.

The heats of neutralization shown in the following table were obtained by mixing 50 cm³ of 2.0 M acid with 50 cm³ of 2.0 M alkali and measuring the rise in temperature of the mixture.

Acid	Alkali	Heat of neutralization (kJ mol^{-1})
Hydrochloric acid, HCl(aq)	Potassium hydroxide, KOH(aq)	−57.9
Hydrochloric acid, HCl(aq)	Sodium hydroxide, NaOH(aq)	−57.9
Nitric acid, HNO₃(aq)	Sodium hydroxide, NaOH(aq)	−57.9
Hydrobromic acid, HBr(aq)	Potassium hydroxide, KOH(aq)	−57.9

(a) Draw a diagram of a simple apparatus you would use to measure these heats of neutralization and label each part.

(b) What measurements would you make
   (i) before mixing the solutions?
   (ii) after mixing the solutions?

(c) Write an ionic equation for the reaction which takes place between hydrochloric acid and potassium hydroxide solutions.

(d) Explain why the heat of neutralization has the same value for all these reactions.

(e) When $50 \, cm^3$ of 2.0 M hydrochloric acid was mixed with $50 \, cm^3$ of 2.0 M ammonia solution, $NH_3(aq)$, there was a temperature rise of 13 °C. Calculate the heat of neutralization for this reaction.　　　　**(L)**

8 Moles per second – reaction rates

8.1 Rates of chemical reactions

In chemical reactions, amounts or concentrations of substances change. Reactants are used up and products are formed. The rates at which these changes take place give a measure of the 'rate of reaction'.

For the reaction

$$Mg(s) + 2HCl(aq) \rightarrow MgCl_2(aq) + H_2(g)$$

we could measure

* the rate of loss of magnesium, or
* the rate of loss of hydrochloric acid, or
* the rate of formation of magnesium chloride, or
* the rate of formation of hydrogen.

In this case it is easiest to follow the formation of hydrogen by measuring its volume with time.

Any property which changes with the amount or concentration of a reactant or product can be followed and then

* $\text{Rate} = \dfrac{\text{Change recorded in any property}}{\text{Time for the change}}$

Rates in chemical reactions may be measured in a variety of practical units but they can be converted to a single, useful unit – mol/s.

Example 8a
In the reaction of magnesium with dilute hydrochloric acid, $48 \, cm^3$ of hydrogen is formed in $10 \, s$ at room temperature and pressure.
(a) What is the average rate of formation of hydrogen (i) in cm^3/s and (ii) in mol/s?
(b) What is the average rate of formation of magnesium chloride in mol/s?

Answer

(a) The average rate of formation of hydrogen $= \dfrac{48 \, cm^3}{10 \, s}$

$= 4.8 \, cm^3/s$

The volume of one mole of hydrogen at room temperature and pressure $= 24\,000 \, cm^3$

So the average rate of formation of hydrogen $= \dfrac{4.8 \, cm^3/s}{24\,000 \, cm^3/mol}$

$= 0.0002 \, mol/s$

(b) From the equation

$$Mg(s) + 2HCl(aq) \rightarrow MgCl_2(aq) + H_2(g)$$

equal amounts of magnesium chloride and hydrogen are formed

So the average rate of formation of magnesium chloride $= 0.0002 \, mol/s$

Questions
1 In example 8a, the mass of magnesium falls from $0.08 \, g$ to $0.032 \, g$ in $10 \, s$.
   (a) What is the average rate of loss of magnesium (i) in g/s and (ii) in mol/s?
   (b) What is the average rate of loss of hydrochloric acid in mol/s?
2 In the reaction of calcium carbonate (marble) with hydrochloric acid, $144 \, cm^3$ of carbon dioxide is formed in $20 \, s$ at room temperature and pressure.

$$CaCO_3(s) + 2HCl(aq) \rightarrow CaCl_2(aq) + H_2O(l) + CO_2(g)$$

   (a) What is the average rate of formation of carbon dioxide (i) in cm^3/s and (ii) in mol/s?
   (b) What is the average rate of loss of calcium carbonate in mol/s?
   (c) What mass of calcium carbonate is used up in $1 \, s$?
3 In the reaction of calcium with water, $0.2 \, g$ of calcium is lost in $100 \, s$ at room temperature and pressure:

$$Ca(s) + 2H_2O(l) \rightarrow Ca(OH)_2(aq) + H_2(g)$$

   (a) What is the average rate of loss of calcium (i) in g/s and (ii) in mol/s?
   (b) What is the average rate of formation of hydrogen (i) in mol/s and (ii) in cm^3/s?

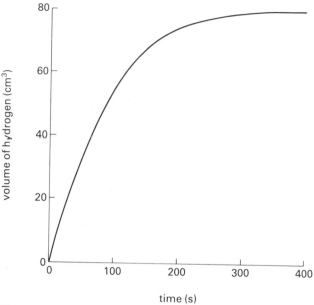

Figure 8.1

In most chemical reactions, the rate changes with time. Figure 8.1 shows the volume of hydrogen formed with time in the reaction of magnesium with dilute hydrochloric acid. The graph is steepest at the start of the experiment, when the reaction is at its fastest. As the reaction continues it becomes slower until eventually it stops. This happens because one of the reactants is being used up.

The most important rate when comparing reactions is the initial rate.

Example 8b

The data for figure 8.1 is given below.

Time (s)	0	50	100	150	200	250	300	350	400
Volume of hydrogen (cm^3)	0	32	55	68	74	78	79	79.5	80

Figure 8.2

What is the average rate of formation of hydrogen, in cm^3/s, over the intervals (a) 0–50 s and (b) 50–100 s?

Answer

(a) Volume of hydrogen formed over the
    interval 0–50 s $= 32\,\text{cm}^3$

    Rate of formation of hydrogen $= \dfrac{32\,\text{cm}^3}{50\,\text{s}}$

                                           $= 0.64\,\text{cm}^3/\text{s}$

(b) Volume of hydrogen formed over the interval 50–100 s

$$= 55\,cm^3 - 32\,cm^3$$
$$= 23\,cm^3$$

Rate of formation of hydrogen

$$= \frac{23\,cm^3}{50\,s}$$
$$= 0.46\,cm^3/s$$

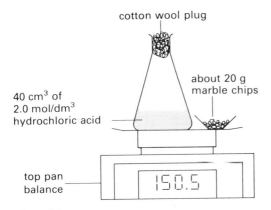

cotton wool plug

about 20 g marble chips

40 cm³ of 2.0 mol/dm³ hydrochloric acid

top pan balance

150.5

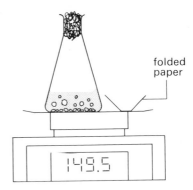

folded paper

149.5

Figure 8.3

Questions

4 (a) Using the data in figure 8.2, calculate the average rates of formation of hydrogen over the intervals (i) 100–150 s, (ii) 150–200 s, (iii) 200–250 s, (iv) 250–300 s, (v) 300–350 s and (vi) 350–400 s.
  (b) Draw a graph of rate (vertical axis) against time.
  (c) How does the rate change with time?
  (d) At what time does the reaction stop?

Questions 5 and 6 are about an investigation of the reaction between marble chips and dilute nitric acid:

$$CaCO_3(s) + 2HNO_3(aq) \rightarrow Ca(NO_3)_2(aq) + H_2O(l) + CO_2(g)$$

The apparatus used is shown in figure 8.3. The results of three experiments, A, B and C are shown in figure 8.4 on the next page.

5 In experiment A, 20 g of small marble chips (an excess) reacts with 40 cm³ of 2.0 mol/dm³ nitric acid. In experiment B, 20 g of large chips are used with 40 cm³ of 2.0 mol/dm³ nitric acid.
  (a) On one set of axes, draw graphs of mass of carbon dioxide formed (vertical axis) against time for experiments A and B.

Time	Mass of carbon dioxide formed (g)		
(s)	Experiment A	Experiment B	Experiment C
30	0.45	0.18	0.22
60	0.85	0.38	0.40
90	1.13	0.47	0.51
120	1.31	0.75	0.58
180	1.48	1.05	0.67
240	1.54	1.25	0.73
300	1.56	1.38	0.76
360	1.58	1.47	0.78
420	1.59	1.53	0.79
480	1.60	1.57	0.79
540	1.60	1.59	0.80
600	1.60	1.60	0.80

Figure 8.4

(b) What is the average rate of formation of carbon dioxide over the first 30 s (i) in experiment A and (ii) in experiment B?

(c) Which reaction has the greater initial rate?

(d) After what time did the reaction stop (i) in experiment A and (ii) in experiment B?

(e) What is the effect of particle size on the rate of reaction?

(f) For a given mass of marble, how is particle size related to surface area?

6   In experiment C, 20 g of small marble chips is reacted with 40 cm^3 of 1.0 mol/dm^3 nitric acid.

(a) Draw a graph of mass of carbon dioxide formed (vertical axis) against time for experiment C. For comparison, include on your graph the results of experiment A.

(b) What is the average rate of formation of carbon dioxide over the first 30 s in experiment C?

(c) Of reactions A and C, which has the greater initial rate?

(d) Approximately how many times faster or slower is reaction A than reaction C?

(e) What is the effect of changing the acid concentration on the initial rate of reaction?

(f) How many moles of nitric acid are used (i) in experiment A and (ii) in experiment C?

(g) Why is the mass of carbon dioxide formed in experiment C half that formed in experiment A?

(h) What final mass of carbon dioxide would be obtained if the experiment were repeated using (i) 80 cm^3 of 1.0 mol/dm^3 nitric acid and (ii) 40 cm^3 of 0.5 mol/dm^3 nitric acid?

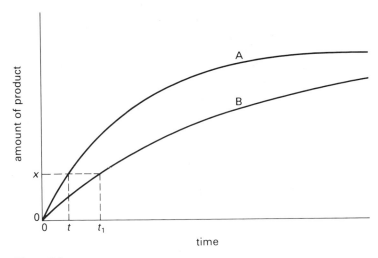

Figure 8.5

In drag racing, cars are timed over a fixed, quarter-mile distance. The fastest car covers the distance in the shortest time. Similarly, the rates of chemical reactions can be compared by finding the time to reach a certain point. In figure 8.5, line A shows the formation of a product under a particular set of conditions. The time, t, is for the formation of a certain amount of product, x. Line B shows a slower reaction. The time, t_1, is for the formation of the same amount of product, x. t_1 is greater than t.

The average rate of formation of product on line A $= \dfrac{x}{t}$ mol/s

The average rate of formation of product on line B $= \dfrac{x}{t_1}$ mol/s

x is the same in both cases, so

the average rate of formation of product on line A $\propto \dfrac{1}{t}$ mol/s

the average rate of formation of product on line B $\propto \dfrac{1}{t_1}$ mol/s

---

Example 8c
In experiment X, 0.5 g of powdered calcium carbonate reacts with 25 cm^3 of 2.0 mol/dm^3 hydrochloric acid in a flask, and carbon dioxide is collected in a gas syringe or measuring cylinder (see figure 8.6). In experiment Y, the acid is diluted to half its original concentration and in experiment Z it is diluted to half again. Figure 8.7 shows the time to collect a certain volume (x cm^3) of carbon dioxide with each concentration of acid.

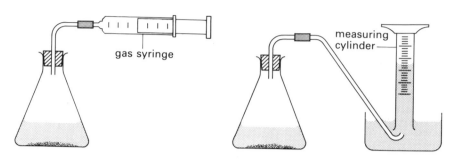

Figure 8.6

	Acid concentration (mol/dm³)	Time to collect x cm³ carbon dioxide, t (s)
Experiment X	2.0	8
Experiment Y	1.0	16
Experiment Z	0.5	32

Figure 8.7

(a) Which is the fastest reaction?
(b) Work out values of $1/t$. $1/t$ is proportional to the initial rate.
(c) Draw a graph of $1/t$ (vertical axis) against concentration.
(d) How is the rate related to the concentration of the acid?

Answer
(a) Experiment X has the fastest initial rate because the volume of carbon dioxide was collected in the shortest time
(b) Figure 8.8 shows values of $1/t$ at each concentration

Acid concentration (mol/dm³)	Time to collect x cm³ carbon dioxide, t (s)	$1/t$ (/s)
2.0	8	0.125
1.0	16	0.062
0.5	32	0.031

Figure 8.8

(c) The graph of $1/t$ against concentration is shown in figure 8.9
(d) The rate is proportional to the concentration. As the concentration increases, the rate increases

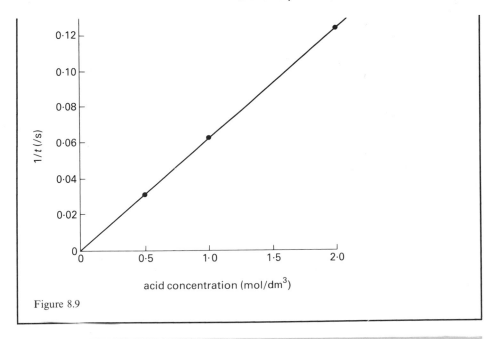

Figure 8.9

Questions

7 5 cm lengths of magnesium ribbon are added to excess dilute hydrochloric acid and the hydrogen is collected in a gas syringe. Figure 8.10 shows the times to collect 10 cm³ of hydrogen using different concentrations of acid, but keeping the total volume of solution constant.

Concentration of acid (mol/dm³)	0.2	0.4	0.6	0.8	1.0
Time to collect 10 cm³ of hydrogen (s)	60	30	20	15	12

Figure 8.10

(a) Draw a graph of time to collect the hydrogen (vertical axis) against concentration of acid.
(b) What effect does increased concentration have on the time to collect 10 cm³ of hydrogen?
(c) How long would it take to collect 10 cm³ of hydrogen with an acid concentration of 0.5 mol/dm³?
(d) Work out values of $1/t$ for each concentration of acid used.
(e) Draw a graph of $1/t$ (vertical axis) against concentration of acid.
(f) How does the rate vary with the concentration of hydrochloric acid?

8 In the reaction of sodium thiosulphate solution with dilute hydrochloric acid, a precipitate of sulphur is formed. The mixture becomes cloudy.

$$Na_2S_2O_3(aq) + 2HCl(aq) \rightarrow 2NaCl(aq) + H_2O(1) + SO_2(aq) + S(s)$$

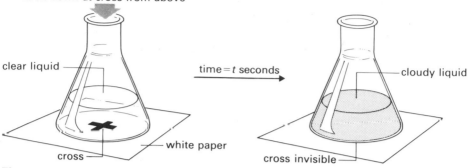

Figure 8.11

In the experiment shown in figure 8.11, the cross becomes obscured when a certain amount of sulphur is formed. The times for the cross to be obscured at different solution temperatures are shown in figure 8.12.

Temperature of mixture (°C)	20	30	40	50	60
Time for cross to be obscured, t (s)	260	130	65	31	16

Figure 8.12

(a) At which temperature does the fastest reaction occur?
(b) Draw a graph of the time for the cross to be obscured, t (vertical axis) against the temperature.
(c) What is the effect on the time, t, as the temperature is raised (i) from 20 °C to 30 °C and (ii) from 30 °C to 40 °C?
(d) Calculate values of $1/t$ at each temperature.
(e) Draw a graph of $1/t$ (vertical axis) against temperature.
(f) What is the effect of increasing temperature on the rate of reaction?
(g) For each 10 °C rise in temperature, by how much, approximately, does the rate increase?

Reaction rates are increased by

* increasing the surface area of a solid reactant
* increasing the concentration of a reactant in solution
* increasing the pressure (if the reactants are gases)
* increasing the temperature
* catalysts.

The *extent* of a reaction is only changed if the amount of reactant not in excess is changed.

Example 8d
Figure 8.13 is a graph of the volume of hydrogen produced when excess
zinc turnings react with 50 cm³ of 2.0 mol/dm³ hydrochloric acid at 20 °C,
against time.

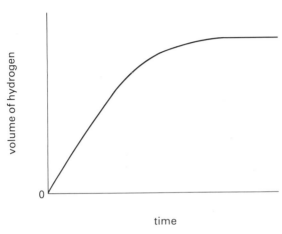

Figure 8.13

What would be the effect of
(a) using 50 cm³ of 1.0 mol/dm³ acid in place of 2.0 mol/dm³ acid?
(b) carrying out the reaction at 30 °C?
(c) adding a few drops of copper(II) sulphate solution to catalyse the
    reaction?
(d) using larger pieces of zinc?

Answer
The effects are shown in figure 8.14 and on the graph in figure 8.15 on the
next page.

	Effect on ...	
	...initial rate	**...extent of reaction**
(a)	Slower – about half the rate in the original reaction	Less – half the amount of acid is used, so half the amount of hydrogen is formed
(b)	Faster – the temperature rise is 10 °C, so the rate is about doubled	Unchanged
(c)	Faster	Unchanged
(d)	Slower – the surface area of zinc is less	Unchanged

Figure 8.14

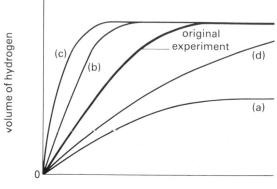

Figure 8.15   time

Questions

9 Excess calcium carbonate in the form of marble chips is reacted with
   25 cm³ of 2.0 mol/dm³ nitric acid at 20 °C. The rate is followed by
   collecting the carbon dioxide formed. What effect on (i) the initial rate
   and (ii) the final volume of carbon dioxide collected, is expected if
   (a) calcium carbonate powder is used in place of marble chips?
   (b) the reaction is carried out at 15 °C?
   (c) 25 cm³ of 4.0 mol/dm³ nitric acid is used?
   (d) 25 cm³ of water is added to the original mixture?

10 Hydrogen peroxide solution decomposes slowly, releasing oxygen:

$$2H_2O_2(aq) \rightarrow 2H_2O(l) + O_2(g)$$

   The reaction is catalysed by manganese(IV) oxide, MnO_2. Figure 8.16
   shows the volume of oxygen released with time when one measure of
   manganese(IV) oxide powder is added to 50 cm³ of hydrogen peroxide
   solution.

Time (s)	0	20	40	60	80	100	120	140	160	180
Volume of oxygen (cm³)	0	10	20	26	32	35	38	39	40	40

Figure 8.16

   (a) Draw a graph of volume of oxygen released (vertical axis) against
       time. Use a vertical scale that goes up to 100 cm³.
   (b) On the same axes, sketch the graphs you would expect if, in separate
       experiments,
        (i) the temperature is raised to 40 °C,
       (ii) 100 cm³ of hydrogen peroxide solution is used,
      (iii) manganese(IV) oxide granules are used in place of powder,
       (iv) 50 cm³ of hydrogen peroxide diluted to half its original
            concentration is used.

11 Calcium reacts with water, giving hydrogen:

$$Ca(s) + 2H_2O(l) \rightarrow Ca(OH)_2(aq) + H_2(g)$$

Figure 8.17 shows some results obtained when 0.2 g of calcium turnings reacts with 100 cm^3 of water at 20 °C.

Time (s)	0	15	30	45	60	75	90	120	150	180
Volume of hydrogen (cm³)	0	7	15	27	39	54	70	94	108	115

Figure 8.17

(a) How many moles of calcium are used?
(b) What volume of hydrogen is released when the calcium reacts completely at room temperature and pressure?
(c) Draw a graph of volume of hydrogen (vertical axis) against time. Show the volume reaching its final volume at about 220 s.
(d) At which of the following times is the rate greatest: 30 s, 90 s, 150 s?
(e) What explanation could account for the slow start to the reaction?
(f) On the axes drawn in (c), sketch the lines you would expect if
    (i) the reaction is repeated, but with the water at 10 °C,
    (ii) the calcium turnings are thoroughly cleaned with emery cloth before use,
    (iii) calcium filings are used in place of turnings.
(g) What should be done to change the extent of reaction?

8.2 Radioactive decay

Radioactive atoms disintegrate and give new atoms. Figure 8.18 on the next page shows the decay of protactinium-234. The activity due to protactinium-234, measured as counts/s, falls with time as the isotope decays away. Each count rate is corrected for the 'background radiation' by subtracting from each observed rate the count rate without any 'active' sample present.

✳ The half-life, $t_{\frac{1}{2}}$, is the time for the activity (or amount) of a radioactive species to fall to half its initial level.
✳ For a particular isotope, the half-life is constant.

Half-lives of isotopes vary from fractions of a second to millions of years. The half-life of an isotope is not affected by a change of concentration or temperature, or by the addition of a catalyst.

Example 8e
Using the experimental results in figure 8.18, find the half-life of protactinium-234.

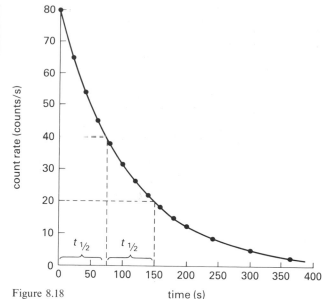

Figure 8.18

Answer

Count rate $= 80$ counts/s at $t = 0\,\text{s}$

Count rate $= \dfrac{80 \text{ counts/s}}{2} = 40$ counts/s at $t = 75\,\text{s}$

Half-life $= 75\,\text{s} - 0\,\text{s}$
$= 75\,\text{s}$

Count rate $= 40$ counts/s at $t = 75\,\text{s}$

Count rate $= \dfrac{40 \text{ counts/s}}{2} = 20$ counts/s at $t = 150\,\text{s}$

Half-life $= 150\,\text{s} - 75\,\text{s}$
$= 75\,\text{s}$

The half-life of protactinium-234, $t_{\frac{1}{2}}$, is $75\,\text{s}$

Questions

12   Figure 8.19 shows count rates corrected for background radiation for a sample containing iodine-128, recorded over a period of 80 minutes.

Time (minutes)	0	5	10	15	20	30	40	50	60	70	80
Count rate (counts/s)	128	110	95	83	72	56	43	32	24	18	14

Figure 8.19

(a) Draw a graph of count rate (vertical axis) against time.

(b) Find the times for the count rate to fall to half its level from three different initial count rates. What is the average of these times?

(c) What is the half-life of iodine-128?

13 Figure 8.20 shows observed count rates, in counts/minute, for a sample containing sodium-24, recorded over several days. The average background radiation count rate was 14 counts/minute.

Time (hours)	0	5	10	15	20	30	40	50	60	70	80
Count rate (counts/min)	2454	1889	1514	1234	989	624	404	254	165	104	76

Figure 8.20

(a) Work out the corrected count rate at each time.

(b) Draw a graph of corrected count rate (vertical axis) against time.

(c) From the graph, work out the half-life of sodium-24.

Example 8f

The half-life of protactinium-234 is 75 s. If the corrected count rate of a sample containing protactinium-234 is 120 counts/s at a certain time,

(a) what is the count rate after 300 s?

(b) how long does it take for the activity to fall to 30 counts/s?

Answer

Figure 8.21 shows how the count rate falls with time

Time (s)	Number of half-lives	Count rate (counts/s)	Fraction of original activity
0		120	
75	1	60	$\frac{1}{2}$
150	2	30	$\frac{1}{4}$
225	3	15	$\frac{1}{8}$
300	4	7.5	$\frac{1}{16}$

Figure 8.21

(a) After 300 s, count rate = 7.5 counts/s

(b) When the activity is 30 counts/s, it is one-quarter of the initial activity. This happens when two half-lives have passed, so time for the activity to fall to 30 counts/s = 150 s

Questions

14 The count rate due to an isotope is 280 counts/minute at a certain time. If the half-life is 30 minutes, what are the count rates after
(a) 30 minutes, (b) 60 minutes and (c) 90 minutes?

15 The half-life of sodium-24 is 15 hours. If the count rate of a sample containing sodium-24 is 160 counts/s,
(a) what will be the count rate after 45 hours?
(b) how long does it take for the count rate to fall to 10 counts/s?

16 The half-life of carbon-14 is 5730 years. How long will it take for a sample of carbon-14 to decay to one-eighth of its present amount?

17 The half-life of iodine-131 is eight days. If the mass of iodine-131 is 0.008 g, what mass will remain after 32 days?

18 The half-life of caesium-137 is 30 years. A sample has an initial mass of 1 g.
(a) Draw a graph to show how the mass of the sample (vertical axis) will change over 200 years.
(b) Approximately what mass will remain in 100 years?
(c) How long will it take for the mass of caesium-137 to become one-fifth of its present mass?

Examination questions

Few practice questions have been included in this chapter. The ideas are straightforward and are frequently tested in public examinations, using a relatively small number of reactions. These include

∗ reaction of a carbonate with a dilute acid
∗ reaction of a metal with a dilute acid
∗ reaction of a reactive metal with water
∗ catalytic decomposition of hydrogen peroxide
∗ reaction of sodium thiosulphate solution with dilute acid.

A wide selection of these public examination questions follows.
  In some of the questions you are asked to write on a graph or complete a table. Do not write in this book but show your answer on a copy of the graph or table drawn on your answer paper.

1 This question is about the rates of chemical reactions.
(a) The rate of the reaction between calcium carbonate and hydrochloric acid was investigated in three experiments.

Experiment A
The following results were obtained when 0.3 g of powdered calcium carbonate and 20 cm^3 of hydrochloric acid were used at 25°C. The acid was in excess.

Volume of gas (cm^3)	0	18	36	54	64	70	72	72
Time (seconds)	0	10	20	30	40	50	60	70

(i) Plot the results of experiment A on a graph. Label this curve A.
(ii) Use your graph to find the volume of gas formed in the first 35 seconds. Use your graph to find how long it took for the volume of gas collected to increase from 40 cm³ to 60 cm³.
(iii) Explain why the reaction stops after about 60 seconds.
(iv) Name the gas given off. Describe the test for this gas.
(v) Name the salt formed in the reaction.
(vi) Complete and balance the following equation for the reaction:

$$CaCO_3 + \qquad HCl \rightarrow$$

(b) *Experiment B*
Experiment A was repeated using the same quantities of powdered calcium carbonate and hydrochloric acid but at 40 °C.
  Draw, on the graph used earlier in the question, the position and shape of the curve you would expect to obtain. Label this curve B.

(c) *Experiment C*
The experiment was repeated again, using marble chips in place of the powdered calcium carbonate. The quantities and temperature were the same as in experiment A.
(i) Would you expect the rate of the reaction in experiment C to be faster, slower or the same as the rate in experiment A?
(ii) Give a reason for your answer.

(d) The rates of some chemical reactions are increased by the presence of light and/or increased pressure.
(i) State *one* example of a reaction affected by light.
(ii) State *one* example of a reaction affected by pressure.   **(EAEB)**

2 50 cm³ of dilute hydrochloric acid was added to an excess of calcium carbonate contained in a flask connected to a graduated syringe. The gas evolved was collected in the syringe at room temperature and pressure and the total volume recorded at 10-second intervals. When no more gas was evolved some calcium carbonate remained unreacted in the flask. The results are shown in the table.

Time (secs)	10	20	30	40	50	60	70	80	90	100	110	120
Total volume (cm³)	130	225	300	360	410		480	490	500	500	500	500

(a) Plot the results on a graph.
(b) Estimate, from the graph, the total volume of gas collected at 60 seconds.
(c) Write an equation for the reaction.
(d) Calculate the concentration, in mol/dm³, of the hydrochloric acid.
(e) On the same graph sketch a second curve, labelled "Expt. 2", to show the results you would expect if the experiment were repeated with the same quantities of material but with the hydrochloric acid at a higher temperature, the gas still being collected at room temperature.   **(C)**

3   The following results were obtained when 50 cm³ of molar (M) hydrochloric acid reacted with 5 g of powdered calcium carbonate and the volume of gas evolved was measured at regular intervals. Temperature was kept constant.

Time (minutes)	1	2	3	4	5	6	7	8
Total volume of gas (cm³)	280	420	490	530	550	560	560	560

(a)   (i) Plot and label a graph of total volume of gas against time.
  (ii) Label the point where the rate of reaction is a maximum.
  (iii) Label the point where the reaction ceases.
  (iv) Explain how, and why, the rate of reaction changes as the reaction proceeds.

(b)   5 g of calcium carbonate reacts with exactly 50 cm³ of 2 M hydrochloric acid.
  (i)   Which reactant was in excess in the above experiment?
  (ii)   On the same axes *sketch*, and label, graphs showing the results that would be expected if the following experiments were carried out under the same conditions.

Experiment I
5 g of a similar sample of calcium carbonate with 100 cm³ of 0.5 M hydrochloric acid.

Experiment II
5 g of a similar sample of calcium carbonate with 50 cm³ of 2 M hydrochloric acid.

(c)   Give the name or formula of each of the three substances formed when calcium carbonate reacts with hydrochloric acid.

(d)   How would the initial rate of reaction be affected by the following changes:
  (i)   an increase in temperature,
  (ii)   an increase in particle size of the calcium carbonate?   (EAEB)

4 (a)   An experiment was carried out using pieces of magnesium ribbon each weighing 0.05 g. These pieces of magnesium ribbon were reacted with solutions of dilute hydrochloric acid of different concentrations. The volume of hydrogen produced was measured at intervals in each case.
  (i)   The first piece of magnesium ribbon was accurately weighed. How could other pieces of equal mass be obtained without using a chemical balance?
  (ii)   Draw a labelled diagram of apparatus that could be used for this experiment.

(b)   A piece of magnesium ribbon was added to 5 cm³ of a hydrochloric acid solution labelled A. The results are plotted in the graph opposite.
  (i)   What volume of hydrogen was produced when the reaction had finished?
  (ii)   How long did it take for the reaction to be completed?

(c)   A piece of magnesium ribbon was added to 5 cm³ of another hydrochloric acid solution labelled B. The results are shown in the table below.

Time (s)	0	5	10	15	20	25
Volume of hydrogen (cm³)	0	25	40	45	50	50

Plot these results on the graph.

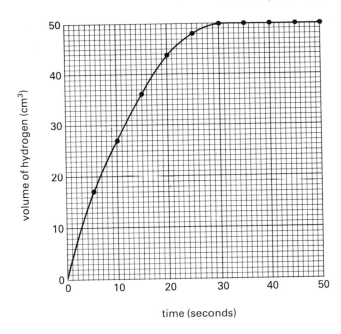

(d) Which of the solutions A or B is a more concentrated solution of hydrochloric acid? Give a reason for your answer.

(e) A third piece of magnesium ribbon was added to a solution of hydrochloric acid labelled C. At the end of the experiment only 40 cm³ of hydrogen had been collected and some magnesium remained in the solution.

   (i) If no hydrogen had leaked from the apparatus, why was less hydrogen collected in this case?

   (ii) What mass of magnesium remained at the end of the experiment?   **(EAEB)**

5 10 cm³ (an excess) of 1 M HCl were added to 0.048 g of magnesium ribbon in a suitable apparatus at 20 °C and atmospheric pressure. The volume of hydrogen formed was noted every 30 seconds, and a graph of volume of gas produced against time from the start of the experiment was plotted as shown on the next page.

(a) At what time was the action most rapid?

(b) How many seconds had elapsed before all the magnesium had dissolved?

(c) What volume of hydrogen had been produced when all the magnesium had dissolved?

(d) How many seconds had elapsed before half of the magnesium had dissolved?

(e) How many moles of atoms are there in 0.048 g of magnesium?

(f) An equation for the reaction is:

$$Mg(s) + 2HCl(aq) \rightarrow MgCl_2(aq) + H_2(g)$$

What volume of 1 M HCl would be required to react completely with 0.048 g of magnesium?

(g) Sketch in, on the graph above, the curve you would expect if the experiment was repeated in an identical manner, except that the temperature was 30 °C. **(L)**

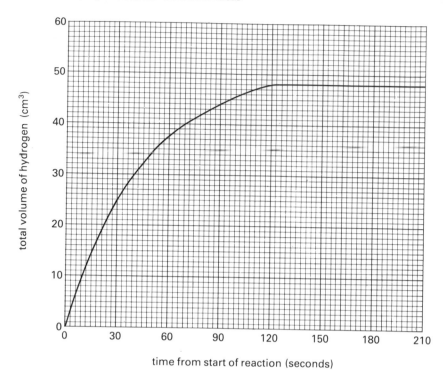

total volume of hydrogen (cm³) vs time from start of reaction (seconds)

6 (a) A length of magnesium ribbon was placed in a measured volume of 2 M hydrochloric acid solution in a beaker. The reaction was complete in two minutes. Write down an equation for the reaction, name the products and describe the observations which would have been made.

   (b) An identical experiment using 2 M ethanoic (acetic) acid was then carried out. How would the rate of this reaction compare with the rate of the reaction using 2 M hydrochloric acid solution? Give an explanation for your answer. **(NISEC)**

7 5 g of granulated zinc are put in a conical flask and covered with 100 cm³ (an excess) of 2 M hydrochloric acid at 20 °C. A fairly slow reaction takes place.

   How would you expect the rate of formation of hydrogen to be affected if each of the following changes is made in turn, all other conditions remaining identical?

New condition	Change, if any	Reason
Using 5 g of powdered zinc		
Using 3 g of granulated zinc		
Using 100 cm³ of 2 M ethanoic (acetic) acid		
Altering the temperature to 40 °C		
Adding a few drops of aqueous copper(II) sulphate		

Simply state in the table whether you would expect the rate to be increased, decreased or unchanged, and in each case give a reason for your answer.   **(L)**

8 In an experiment to study the rate of reaction between zinc and hydrochloric acid, 0.2 g of zinc and 250 cm³ of 1.0 M hydrochloric acid (an excess) were reacted together. The volume of hydrogen formed was measured at suitable time intervals by using a gas syringe attached to the reaction flask. Room temperature was 20 °C. The zinc dissolves in the acid during the reaction; the reaction may be represented as:

$$Zn(s) + 2H^+(aq) \rightarrow Zn^{2+}(aq) + H_2(g)$$

Results are given in the form of the graph below.

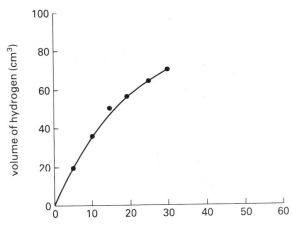

(a) Calculate the volume of gas that the syringe should contain when the reaction is complete.

(b) Sketch on the graph the way in which the line might continue if the reaction rate had been measured for 60 minutes.

(c) Sketch on the graph the line you would expect to obtain by repeating the experiment but this time keeping the reaction flask at 15 °C. Clearly label this line C.

(d) Copper powder added to the reaction mixture catalyses the reaction. In a repeat of the experiment at 20 °C with 0.2 g of copper powder present, when reaction was complete
  (i) what volume of gas would be expected?
  (ii) how would you recover copper from the mixture in the flask?

(e) Sketch on the graph the line you would expect to obtain in following the reaction of 0.2 g of zinc with 250 cm³ of 1.0 M sulphuric acid (H_2SO_4) at 20 °C for 60 minutes. Clearly label this line E.

(f) In these experiments, the gas collected in the syringe is a mixture of air and hydrogen. Why may this volume reading be plotted on the graph as a volume of hydrogen?

(g) If a portion of the liquid obtained at the end of the reaction with sulphuric acid were transferred to a watch glass and left on the bench, what would you expect to see after a day or so and how do you account for this?   **(L)**

9 A stick of lithium was removed from a bottle of oil in which it was stored and a piece was cut from it which was found to have a mass of 0.1 g. The piece was

transferred to an apparatus containing an excess of water. After the reaction had started, the volume of hydrogen produced was measured at intervals. Some of the results are shown in the table below.

Time (minutes)	Volume of hydrogen (cm³)
1	8
2	24
3	72
4	138
5	172
6	172

(a) Draw a diagram of a suitable apparatus for the experiment in which you could ensure that the lithium did not react until you were ready to start the reaction.

(b) Plot the results on suitably labelled axes, with time on the horizontal axis.

(c) What is the rate of the reaction in volume of hydrogen per minute between 3 and 4 minutes?

(d) Explain precisely the reasons for the change in rate between 4 and 5 minutes.

(e) It was thought that the results in the first 4 minutes were affected by oil on the lithium. So the experiment was repeated with 0.1 g of lithium from which all oil had been removed. Draw a *dotted* line on the graph to show the results you would expect from this second experiment.

(f) (i) When 1 mole of lithium has completely reacted, what volume of hydrogen would be produced at room temperature and pressure?

(ii) Lithium hydroxide, LiOH, is also produced. Use the results above to complete and balance the following equation:

$$Li(s) + H_2O(l) \rightarrow LiOH(aq)$$

(g) Give *one* property of the resulting solution.   (L)

10 0.1 g of calcium turnings was reacted with 25 cm³ (an excess) of cold water at room temperature. The volume of hydrogen given off was measured every 30 seconds for 4 minutes. The results were:

Time (seconds)	0	30	60	90	120	150	180	210	240
Volume (cm³)	0	20	32	42	50	56	59	60	60

(a) What is meant by 'excess water'?

(b) Plot the results on graph paper. (Scales: horizontal axis, 2 cm represents 30 seconds; vertical axis, 2 cm represents 10 cm³.) Label the curve Graph A.

(c) (i) After how many seconds did the reaction stop?

(ii) Why did the reaction finish at that time?

(d) What does the shape of the graph indicate about changes in the speed of the reaction?

(e) (i) What would happen to the rate of the reaction and the total volume of hydrogen finally produced if 0.05 g of calcium turnings was used with 25 cm³ of water at room temperature?

(ii) On the same axes as the original Graph A, sketch the curve you would obtain by using 0.05 g of calcium turnings and 25 cm³ of water at room temperature. Label the curve Graph B.   (NWREB)

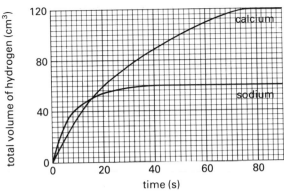

11 The curves in the graph above show how the total volume of hydrogen evolved
changed with time when P g of sodium and Q g of calcium were added to separate
large volumes of water at room temperature and pressure.
    (Under the conditions of the experiment 1 mol of hydrogen gas occupies $24\,dm^3$.)
(a) Write an equation for the reaction of water with (i) sodium and (ii) calcium.
(b) Calculate the values of P and Q.
(c) At what time was (i) half the mass of sodium and (ii) half the mass of calcium,
     used up?
(d) In each case, name *one* metal that docs react with water but
     (i) initially more rapidly than sodium,
     (ii) initially more slowly than calcium.
(e) If P g of sodium had been added to an excess of ethanol instead of water in the
     above experiment, how, if at all, would each of the following alter:
     (i) the initial rate of production of hydrogen,
     (ii) the final volume of hydrogen produced?
     Give reasons for your answer in (ii).   (C)

12 The decomposition of hydrogen peroxide

$$2H_2O_2 \rightarrow 2H_2O + O_2$$

is catalysed by manganese(IV) oxide. The rate of reaction can be followed by
measuring the volume of oxygen produced at intervals over a period of time.
(a) Sketch the apparatus you would use to do this experiment.
(b) The following data were obtained in an experiment using a solution of
     hydrogen peroxide and a powdered catalyst.

Time (s)	5	10	15	20	25	30	35	40	45	50
Volume of oxygen measured at $25\,°C$ and 1 atmosphere (cm^3)	25	39	47	53	57	60	62	63	63	63

    Use squared paper to plot a graph of volume against time.
(c) What volume of gas would have been obtained on completion of the reaction if
     the volume had been measured at $15\,°C$ and 1 atmosphere?
(d) How many moles of oxygen would have been produced at (i) $0\,°C$, (ii) $25\,°C$?
(e) On your graph sketch the shape of the curve obtained using the same mass of a
     coarsely-grained catalyst rather than the fine powder used originally. State the
     volume of oxygen obtained at $25\,°C$.

(f) How many moles of hydrogen peroxide were used?

(g) What volume of 0.5 M hydrogen peroxide does this represent?   (O)

13 The equation for the decomposition of hydrogen peroxide is:

$$2H_2O_2(aq) \rightarrow 2H_2O(l) + O_2(g)$$

The reaction is very slow but is catalysed by a wide range of substances including manganese(IV) oxide.

A solution of hydrogen peroxide should be stored in a dark glass bottle with a rubber stopper with a small slit in it.

(a) Name the gas produced on the decomposition of hydrogen peroxide.

(b) Why is the solution of hydrogen peroxide stored in a dark glass bottle?

(c) Why is a rubber stopper with a slit preferable to a screw top for a bottle of hydrogen peroxide?

(d) An experiment was carried out using the apparatus below.

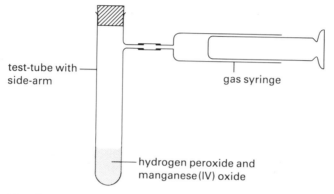

test-tube with side-arm

gas syringe

hydrogen peroxide and manganese(IV) oxide

10 cm³ of hydrogen peroxide solution was placed in the test-tube and 0.5 g of manganese(IV) oxide added. The rubber bung was quickly replaced and the volume of gas collected was measured at 30 second intervals. The results are shown in the table below.

Time seconds)	0	30	60	90	120	150	180	210
Volume of gas collected (cm³)	0	27	45	57	64	68	70	70

(i)   Plot a graph of the volume of gas collected against time using the results in the table.

(ii) After how long was the reaction complete?

(iii) What would be the final volume of gas collected if no manganese(IV) oxide had been added?

(iv) What mass of manganese(IV) oxide would remain at the end of the experiment?

(v)   The concentration of a hydrogen peroxide solution is often expressed in terms of 'volume strength'. A solution of hydrogen peroxide is said to be '20 volume' if on decomposition 1 cm³ of this solution produces 20 cm³ of gas.

What is the volume strength of the hydrogen peroxide used in this experiment?   (EAEB)

14 The concentration of solutions of hydrogen peroxide in water can be determined by measuring the volume of oxygen given off when the hydrogen peroxide, in a fixed volume of solution, is completely decomposed. For example, one litre of '20 volume' hydrogen peroxide, after complete decomposition, gives 20 litres of oxygen at room temperature and pressure. This decomposition can be represented by the following equation:

$$2H_2O_2(aq) \rightarrow 2H_2O(l) + O_2(g)$$

Manganese(IV) oxide is a suitable catalyst for this reaction.
(a) Show, on a graph, the shapes of the curves you would expect to obtain when
  (i) $10\,cm^3$ of '20 volume' hydrogen peroxide decomposes completely in the presence of the catalyst.
  (ii) $20\,cm^3$ of '10 volume' hydrogen peroxide is used under the same experimental conditions as in (a)(i).
  Label your curves (i) and (ii) to correspond with your answers to parts (a)(i) and (a)(ii).
(b) Calculate the number of moles of hydrogen peroxide in one litre of '20 volume' solution.
(c) It was observed that when a mixture of hydrogen peroxide and manganese(IV) oxide was shaken in a flask, gas was given off rapidly but, when the catalyst settled to the bottom of the flask, the gas came off more slowly. How do you account for this?
(d) Copper(II) oxide is also a catalyst for the decomposition of hydrogen peroxide. How would you attempt to compare the effect of copper(II) oxide in this reaction with that of manganese(IV) oxide?
  (i) Describe briefly *or* draw a diagram of the apparatus you would use.
  (ii) State *three* conditions that must be fulfilled before the experiments may be regarded as a true comparison.
(e) If copper(II) oxide is found to be a less active catalyst than manganese(IV) oxide, what can you say about the volume of oxygen formed when $10\,cm^3$ of '20 volume' hydrogen peroxide is completely decomposed by copper(II) oxide? Give a reason to support your answer.
(f) In addition to manganese(IV) oxide and copper(II) oxide, chromium(III) oxide and iron(III) oxide catalyse the decomposition of hydrogen peroxide. What have these oxides in common? **(L)**

15 In an experiment, 0.5 g of manganese(IV) oxide was added to $50\,cm^3$ of 0.2 M hydrogen peroxide and the volume of oxygen evolved was measured at regular time intervals. The results of the experiment are shown in the graph on the next page.
(a) Draw a sketch of the apparatus you would use to carry out the experiment.
(b) Use the graph to find the volumes of oxygen evolved
  (i) during the first 5 minutes (from 0 to 5 min),
  (ii) during the second 5 minutes (from 5 to 10 min),
  (iii) during the fifth 5 minutes (from 20 to 25 min).
  Explain why the volumes produced in 5 minutes vary in this way.
(c) Using the same scales and axes, make a copy of the graph: show clearly on your copy the approximate results you would have expected if the experiment had been carried out under the same laboratory conditions using
  (i) $100\,cm^3$ of 0.1 M hydrogen peroxide and 0.5 g of manganese(IV) oxide.
  (ii) $50\,cm^3$ of 0.4 M hydrogen peroxide and 0.5 g of manganese(IV) oxide. **(C)**

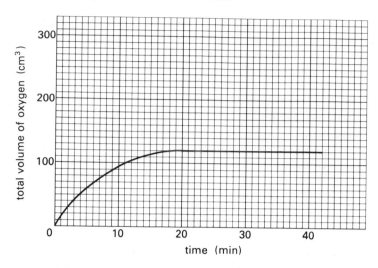

16 A series of experiments was carried out to examine the effects of changing the concentration of hydrochloric acid in the following reaction:

$$Na_2S_2O_3(aq) + 2HCl(aq) \rightarrow 2NaCl(aq) + SO_2(g) + S(s) + H_2O(l)$$

Measured volumes of water, sodium thiosulphate solution and hydrochloric acid were poured into a $100\,cm^3$ beaker. The time was taken for the solution to turn so cloudy that a cross underneath the beaker just disappeared from view.

The results are shown in the table below.

Experiment number	Volume of water (cm³)	Volume of sodium thiosulphate solution (cm³)	Volume of hydrochloric acid (cm³)	Time (s)
1	20	40	5	136
2	15	40	10	74
3	10	40	15	48
4	5	40	20	32

All experiments were carried out at $20\,^{\circ}C$.

(a) What are the meanings of the following state symbols in the equation: (g), (l), (aq)?

(b) Name the substance produced which caused the solution to turn cloudy.

(c) What was the total volume of solution in each experiment?

(d) When should the timing be started?

(e) In which experiment is the reaction fastest?

(f) Plot a graph of the volume of acid used against the time taken for the cross to disappear.

(g) An additional experiment was carried out using $12\,cm^3$ of hydrochloric acid.
   (i) What volume of water should be added?
   (ii) From your graph, how long should it take for the cross to disappear?

(h) What can be concluded from these experiments? Complete the following sentence: 'As the concentration of hydrochloric acid increased...'

(i) Experiment 1 was repeated under the same conditions but using a $250 \, cm^3$ beaker in place of the $100 \, cm^3$ beaker. How would this change the result obtained?

(j) Experiment 1 was repeated again using a $100 \, cm^3$ beaker but at a temperature of $30 \, °C$. How would this change the result obtained? **(EAEB)**

17 The effect of temperature on the rate of reaction between $10 \, cm^3$ of 0.2 M aqueous sodium thiosulphate(VI) and $5 \, cm^3$ of 2 M hydrochloric acid (excess) was studied by noting the time taken for a cross, drawn on a piece of paper underneath the $100 \, cm^3$ conical reaction flask, to disappear at different temperatures. The results are shown in the table below.

Starting temperature (°C)	19	28	38	47	57
Time (seconds)	34	26	18	10	6

(a) Plot a graph of time against temperature and use your graph to estimate the reaction time at (i) $15 \, °C$, (ii) $42 \, °C$.

(b) State and explain the effect upon the rate of reaction at a given temperature of changing the concentration of the sodium thiosulphate(VI) solution.

(c) The cross disappears because of the formation of a yellow suspension of sulphur:

$$Na_2S_2O_3 + 2HCl \rightarrow 2NaCl + S(s) + SO_2 + H_2O$$

Calculate the mass of sulphur produced when the reaction is complete, starting with $10 \, cm^3$ of 0.20 M aqueous sodium thiosulphate(VI).

(d) Explain the apparent effect of carrying out the reaction in a narrower $25 \, cm^3$ conical flask.

(e) State *three other* factors that may affect the rate of chemical reactions, giving the equations of reactions to illustrate each. (You should use a different reaction for each factor, excluding the thiosulphate/hydrochloric acid reaction already described.) **(O)**

18 This question is about measuring the rate of the reaction between solutions of potassium peroxodisulphate, $K_2S_2O_8$, and potassium iodide, KI. It was studied by the following method:

$20 \, cm^3$ of potassium peroxodisulphate solution was mixed with $20 \, cm^3$ of potassium iodide solution (an excess), $5 \, cm^3$ of sodium thiosulphate solution and $1 \, cm^3$ of starch solution.

The potassium peroxodisulphate reacts with the potassium iodide at a measurable rate, forming iodine. As soon as the iodine is formed it reacts rapidly with the sodium thiosulphate until all the sodium thiosulphate is used up. The iodine formed after the sodium thiosulphate is used up gives a blue colour with the starch. The time taken for the blue colour to appear was recorded.

The experiment was repeated using potassium peroxodisulphate solutions of different concentrations and the results are shown in the table overleaf.

(a) If we wish to investigate the rate of a reaction why do we calculate $1/t$?

(b) Plot a graph showing how $1/t$ varies with the concentration of the potassium peroxodisulphate solution.

(c) Explain why this graph shows that the rate of the reaction is proportional to the concentration of the potassium peroxodisulphate solution.

Peroxodisulphate concentration (mol dm^{-3})	Time (t) for blue colour to appear (sec)	$1/t$ (sec^{-1})
0.3	20	0.05
0.2	30	0.033
0.15	40	0.025
0.05	120	0.0083

(d) Use your graph to determine the time that would have been taken for the blue colour to appear if the potassium peroxodisulphate solution had been 0.08 M.

(e) In all the above experiments the solution of sodium thiosulphate used was 0.01 M. What would have been the time taken for the blue colour to appear if the experiment was repeated using 0.2 M potassium peroxodisulphate solution and 0.02 M sodium thiosulphate solution in the same volumes as before?   (L)

19 A series of reactions was carried out to examine the effect of changes in concentration on the reaction between glycerol and potassium permanganate solution at room temperature (20 °C). Solutions of glycerol in water of different concentrations were prepared.

10 cm^3 of one of the glycerol solutions was placed in a test-tube. 5 cm^3 of potassium permanganate solution was added and the clock started. The clock was stopped when the colour of the solution disappeared. The procedure was repeated with the other glycerol solutions. The results are shown in the table below.

Percentage concentration of glycerol	Time for reaction (seconds)
5	300
10	150
15	100
20	75
25	60

(a) Plot a graph of concentration of glycerol against time.

(b) From the graph, find the time required for the reaction if a 12% solution of glycerol had been used.

(c) What can be concluded about the change in the rate of reaction as the concentration of glycerol was increased?

(d) What change would you expect in the results, if the experiment was carried out at 30 °C?

(e) The experiments were repeated with the addition of a few drops of a colourless liquid before the addition of potassium permanganate. The results in the table below were obtained.

Percentage concentration of glycerol	Time for reaction (seconds)
5	252
10	132
15	84
20	60
25	48

    (i) What effect does the colourless liquid have on the time for each reaction?

    (ii) What is the action of the colourless liquid?   **(EAEB)**

20 The decay of a sample of a radioactive isotope was studied using a suitable Geiger-Müller tube and scaler. Readings of scaler counts against time were plotted to give the graph shown below. A count of background radiation in the laboratory was made in a separate experiment and found to be 10 counts per minute on average.

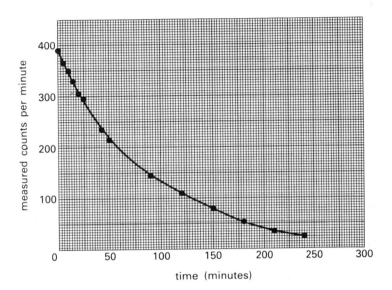

(a) At first, readings were taken every five minutes, but towards the end of the experiment they were taken every thirty minutes. Suggest a reason for this.

(b) It is usual to wear gloves when working with radioactive materials. What danger do gloves give protection against? State one danger they do not give protection against.

(c) After what time would you consider this particular sample to be radioactively harmless? Give reasons to support your answer.

(d) From the decay curve, find the value of the half-life of this isotope.

(e) What mass of a radioactive isotope having a half-life of 30 minutes would remain after 90 minutes if the initial mass of the isotope was 0.010 g?

(f) Uranium-238 decays by alpha emission to form an element X which decays by beta emission to form Y. The element Y decays by beta emission to form Z.

    (i) Complete the decay series relationships by writing the mass number and atomic number of X, Y and Z:

$$^{238}_{92}U \rightarrow X + ^{4}_{2}He$$
$$X \rightarrow Y + ^{0}_{-1}e$$
$$Y \rightarrow Z + ^{0}_{-1}e$$

    (ii) What is the relationship between Z and $^{238}_{92}U$?   **(L)**

21 The radioactive decay series of the elements A, B, C, D, E, F, G and H is shown below and the type of radiation emitted at some stages of the series is indicated above the arrows.

$$^{232}_{90}A \longrightarrow B \xrightarrow{\beta} ^{228}_{89}C \longrightarrow ^{228}_{90}D \xrightarrow{\alpha} E \longrightarrow ^{220}_{86}F \xrightarrow{\alpha} G \xrightarrow{\alpha} H \xrightarrow{\beta}$$

(a) (i) Write down the atomic numbers and mass numbers of elements B and G.
(ii) What name is given to elements such as A and D which have the same atomic number?

(b) By what type of emission does element E decay?

(c) Element H decays by β emission with a half-life of about 10 hours.
   (i) At the beginning of an experiment to plot the decay curve of a sample of element H, the count was 3000 per minute. What count would be expected after 30 hours?
   (ii) Use the value of the half-life for element H given above to plot a decay curve for H. Assume that at zero time there are 3000 counts. Five points plotted will be sufficient for a reasonable curve.
   (iii) From the graph obtained, give the value of the count after 15 hours.

(d) If element G is in Group 6 of the periodic table, in which group would you place element H?   (L)

22 An accident at a nuclear power station releases 1.2 kg of strontium-90 into the atmosphere. If the half-life of this isotope is 28 years, draw a graph to show how the mass of strontium-90 will change over the next 120 years. From your graph predict the amount of Sr-90 remaining (if any) after (a) 50 years, (b) 75 years.   (O)

23 Potassium exists as a radioactive isotope $^{40}_{19}K$ as well as the non-radioactive isotope $^{39}_{19}K$.

(a) Explain the meaning of *isotopes* and state how the potassium isotopes differ from each other in atomic structure.

(b) The product of radioactive decay of $^{40}_{19}K$ is $^{40}_{20}Ca$. State with reasons, the type of radioactive decay that occurs.

(c) If the half-life of the radioisotope $^{40}_{19}K$ is 1.3×10^9 years, calculate, showing your working, how long it will take for 4 g of it to decay to 1 g.   (WJEC)

Multiple choice test

Calculations are frequently included in multiple choice tests. The following multiple choice items have been selected from public examinations and cover many of the ideas in this book.

Each question is followed by five possible answers, **A** to **E**. Select the *one* correct answer in each case. Calculators should not be used. Each question includes all the necessary data.

1 Chromium has an atomic number of 24 and a relative atomic mass of 52. How many neutrons are contained in an atom of chromium?
   A   12
   B   24
   C   26
   D   28
   E   52   (L)

2 How many protons, neutrons and electrons are there in an atom represented by the symbol $^{35}_{17}Cl$?

	Protons	Neutrons	Electrons
A	17	18	17
B	35	17	35
C	17	35	35
D	35	17	17
E	18	17	18	(L)

3 What mass of calcium contains the same number of atoms as 11 g of manganese?
   (Ca = 40, Mn = 55)
   A   11 g

   B $\frac{55}{11} \times 40\,g$

   C $\frac{11}{55} \times 40\,g$

   D $\frac{11}{40} \times 55\,g$

   E   40 g   (L)

4 Which one of the following contains the greatest number of moles of atoms of the element?
   A   3.0 g of carbon, relative atomic mass = 12
   B   4.0 g of aluminium, relative atomic mass = 27
   C   5.0 g of calcium, relative atomic mass = 40
   D   7.0 g of iron, relative atomic mass = 56
   E   8.0 g of cadmium, relative atomic mass = 112   (L)

5 Which of the following volumes of oxygen at room temperature and pressure contains the same number of molecules as 2 g of hydrogen?
   (H = 1, O = 16; 1 mole of molecules of a gas occupies 24 litres at room temperature and pressure)

   A $\frac{2}{24}$ litres

   B $\frac{24}{32}$ litres

   C $\frac{24}{16}$ litres

   D   12 litres

   E   24 litres   (L)

6   How many moles of hydrogen gas contain the same number of atoms as there are in 16 g of helium?
  (He = 4)
  A   1
  B   2
  C   4
  D   8
  E   16   **(NISEC)**

7   Which element is present in the highest proportion by mass in a compound of formula $C_6H_3N_2O_4Cl$?
  (H = 1, C = 12, N = 14, O = 16, Cl = 35.5)
  A   carbon
  B   hydrogen
  C   nitrogen
  D   oxygen
  E   chlorine   **(NISEC)**

8   A metal M forms a chloride MCl_2 of relative formula mass 95. It reacts with aqueous sodium hydroxide to give a precipitate of the metal hydroxide. What is the relative formula mass of the hydroxide?
  (H = 1, O = 16, Cl = 35.5)
  A   24
  B   40
  C   41
  D   58
  E   95   **(NISEC)**

9   A syringe, when full of oxygen gas (O_2) at room temperature and pressure, contained 0.128 g of the gas. The syringe was emptied and refilled with the same volume of a different gas under the same conditions of temperature and pressure. The mass of gas in the syringe this time was 0.064 g. Which of the following gases filled the syringe in the second experiment?
  (H = 1, C = 12, N = 14, O = 16)
  A N_2
  B CO_2
  C NH_3
  D CH_4
  E H_2 **(L)**

10   5.4 g of an element X reacts with oxygen to form 10.2 g of an oxide X_2O_3.
  (O = 16)
  The relative atomic mass of X is
  A   5.4
  B   10.2
  C   15.6
  D   27.0
  E   54.0   **(L)**

11   Phosphorus and sulphur form a compound, 22.2 g of which contains 16.0 g of sulphur. What is the simplest formula of the compound?
  (S = 32, P = 31)
  A   PS
  B P_2S
  C PS_2
  D P_2S_3
  E P_2S_5 **(L)**

12   Experiment shows that 1.2 g of magnesium reacts with 8.0 g of bromine to give 9.2 g of magnesium bromide.
  How much magnesium bromide will be formed if 1.2 g of magnesium is treated with 6.0 g of bromine?
  A   0.9 g
  B   4.6 g
  C   6.9 g
  D   7.2 g
  E   9.2 g   **(L)**

13   A metal M forms oxides of formulae MO, M_2O_3 and MO_3.
  The mass of oxygen which combines with 1 mole of M in each oxide was calculated. These masses would be in the proportion respectively of
  A   1:2:1
  B   1:3:3
  C   1:6:3
  D   2:3:3
  E   2:3:6   **(L)**

14 31 g of a compound contain 12 g of carbon, 3 g of hydrogen and 16 g of oxygen. The relative molecular mass of the compound is 62.
($C = 12$, $H = 1$, $O = 16$)
The molecular formula of the compound is
A CH_3O
B CH_2O_3
C $C_2H_6O_2$
D $C_3H_{10}O$
E $C_{12}H_3O_{16}$ (L)

15 On heating 100 g of a hydrated salt 64 g of the anhydrous salt are formed. If the relative molecular mass of the hydrated salt is 250 and that for water is 18 what is the number of moles of water of hydration in one mole of the hydrated salt?
A 2
B 3
C 4
D 5
E 10 (NISEC)

16 Which of the following fertilizers contains the greatest percentage of nitrogen?
($N = 14$)

	Name	Formula	Mass of 1 mole (grams)
A	Ammonium nitrate	NH_4NO_3	80
B	Ammonium sulphate	$(NH_4)_2SO_4$	132
C	Calcium cyanamide	$CaCN_2$	80
D	Sodium nitrate	$NaNO_3$	85
E	Urea	$CO(NH_2)_2$	60 (L)

17 What is the maximum mass of iron, in grams, which can be extracted from 80 g of iron(III) oxide, Fe_2O_3?
($Fe = 56$, $O = 16$)
A 28
B 32
C 40
D 48
E 56 (L)

18 Some magnesium carbonate on thermal decomposition yielded 0.25 mole of carbon dioxide. What mass of magnesium oxide was produced?
($Mg = 24$, $O = 16$)
A 40 g
B 24 g
C 16 g
D 10 g
E 4 g (NISEC)

19 Each of the copper compounds A to E is reduced to metallic copper when heated in a stream of hydrogen. If 10 g of each compound were separately reduced to the metal, which would give the greatest mass of copper?

	Formula	Mass of 1 mole
A	CuO	80 g
B	$Cu(OH)_2$	98 g
C	$CuCO_3$	124 g
D	CuC_2O_4	152 g
E	$Cu(NO_3)_2$	188 g (L)

20 If 0.7 g of a metal X (relative atomic mass 56) displaces 2.6 g of metal Y (relative atomic mass 208) from a solution of a salt of the latter, the number of moles of atoms of Y displaced by 4 moles of atoms of X is
A 1
B 2
C 4
D 8
E 16 (L)

21 On complete combustion in oxygen 0.1 mole of a hydrocarbon produced 0.3 mole of carbon dioxide and 0.4 mole of water. The molecular formula of the hydrocarbon is

A CH_4

B C_2H_4

C C_2H_6

D C_3H_6

E C_3H_8 **(NISEC)**

22 On complete oxidation 0.01 mole of a hydrocarbon yielded 2.64 g of carbon dioxide and 1.26 g of water.

($CO_2 = 44$, $H_2O = 18$)

The molecular formula of the hydrocarbon is

A C_2H_4

B C_3H_8

C C_6H_6

D C_6H_{14}

E $C_{12}H_{26}$ **(L)**

23 The following compounds burn in air to form carbon dioxide and water only. If 1 g of each compound is burned, which will produce the greatest mass of water?

	Formula	Relative molecular mass
A	C_2H_6	30
B	C_2H_6O	46
C	C_3H_6O	58
D	C_6H_6	78
E	C_6H_6O	94

           (L)

24 What mass of water is formed when 0.72 g of pentane, C_5H_{12}, is burned in excess oxygen?

($H = 1$, $C = 12$, $O = 16$)

A 0.12 g

B 0.18 g

C 0.54 g

D 1.08 g

E 2.16 g   **(L)**

25 What volume of oxygen is used when $20\,cm^3$ of propane (C_3H_8) are burnt completely in oxygen, assuming that the gas volumes are measured under the same conditions of temperature and pressure?

A $60\ cm^3$

B $80\ cm^3$

C $100\ cm^3$

D $120\ cm^3$

E $140\ cm^3$ **(L)**

26 A sample of gas occupied $0.2\,dm^3$ at 18 °C and 1.4 atmosphere pressure. What volume will it occupy at standard temperature and pressure?

A $0.2 \times \dfrac{18}{273} \times \dfrac{1}{1.4}\,dm^3$

B $0.2 \times \dfrac{273}{18} \times 1.4\,dm^3$

C $0.2 \times \dfrac{291}{273} \times 1.4\,dm^3$

D $0.2 \times \dfrac{291}{273} \times \dfrac{1}{1.4}\,dm^3$

E $0.2 \times \dfrac{273}{291} \times 1.4\,dm^3$ **(NISEC)**

27 What volume of hydrogen is produced, at room temperature and pressure, when 0.6 g of magnesium reacts with excess 1 M sulphuric acid?

$$Mg(s) + 2H^+(aq)$$
$$\rightarrow Mg^{2+}(aq) + H_2(g)$$

($Mg = 24$; 1 mole of any gas occupies $24\,000\ cm^3$ at room temperature and pressure)

A $240\ cm^3$

B $600\ cm^3$

C $1000\ cm^3$

D $14\ 400\ cm^3$

E $24\ 000\ cm^3$ **(L)**

28 $100 \, cm^3$ of gaseous ammonia (NH_3) at room temperature and pressure is passed slowly over an excess of heated copper(II) oxide. The equation for the reaction is:

$$2NH_3(g) + 3CuO(s)$$
$$\rightarrow N_2(g) + 3H_2O(l) + 3Cu(s)$$

The volume, in cm^3, of the gaseous product at room temperature and pressure would be

A   25
B   50
C   75
D   100
E   200   (L)

29 $25 \, cm^3$ of dinitrogen oxide (N_2O) is heated with iron wire. The only gaseous product is nitrogen. What volume of nitrogen under the same conditions of temperature and pressure would be formed?

A $12.5 \, cm^3$
B $16.7 \, cm^3$
C $25 \, cm^3$
D $50 \, cm^3$
E $75 \, cm^3$ (L)

30 $60 \, cm^3$ of a gaseous compound of nitrogen and oxygen was completely decomposed into its elements:

$$2N_xO_y(g) \rightarrow xN_2(g) + yO_2(g)$$

The total volume of gas produced was $90 \, cm^3$, which was reduced to $60 \, cm^3$ after passage over heated copper. All volumes were measured at room temperature and pressure. The formula of the gas is

A NO
B NO_2
C N_2O
D N_2O_3
E N_2O_4 (L)

31 $100 \, cm^3$ of nitrogen oxide gas (NO) combine with $50 \, cm^3$ of oxygen to form $100 \, cm^3$ of a single gaseous compound, all volumes being measured at the same temperature and pressure. Which of the following equations fits these facts?

A $NO(g) + O_2(g) \rightarrow NO_3(g)$
B $NO(g) + 2O_2(g) \rightarrow NO_5(g)$
C $2NO(g) + O_2(g) \rightarrow 2NO_2(g)$
D $2NO(g) + O_2(g) \rightarrow N_2O_4(g)$
E $2NO(g) + 2O_2(g) \rightarrow N_2O_6(g)$ (L)

32 What mass of calcium chloride ($CaCl_2$) is required to prepare $500 \, cm^3$ of a $1.0 \, M$ solution? ($Ca = 40$, $Cl = 35.5$)

A $0.50 \, g$
B $1.00 \, g$
C $37.75 \, g$
D $55.5 \, g$
E $500 \, g$ (L)

33 Which of the following solutions contains the *least* number of sodium ions?

A   0.5 litres of $1.0 \, M$ $NaCl$
B   1.0 litres of $0.4 \, M$ $NaOH$
C   1.5 litres of $0.2 \, M$ Na_2SO_4
D   0.2 litres of $0.5 \, M$ Na_2CO_3
E   0.1 litres of $1.0 \, M$ Na_3PO_4 (L)

34 Aqueous sodium hydroxide reacts with a certain metal chloride solution (MCl_n) to form a precipitate of the metal hydroxide:

$$MCl_n + nNaOH$$
$$\rightarrow M(OH)_n + nNaCl$$

$10 \, cm^3$ of $3.0 \, M$ $NaOH$ were found to react exactly with $10 \, cm^3$ of $1.5 \, M$ MCl_n.

The formula of the metal chloride could be

A MCl
B MCl_2
C MCl_3
D M_2Cl
E M_2Cl_3 (L)

35 The reaction between sulphur dioxide and aqueous sodium hydroxide can be represented by the equation

$$SO_2(aq) + 2NaOH(aq)$$
$$\rightarrow Na_2SO_3(aq) + H_2O(l)$$

It was found that $25\,cm^3$ of a saturated solution of sulphur dioxide reacted with $50\,cm^3$ of 2.0 M sodium hydroxide solution. The molarity of the sulphur dioxide solution is

A   0.5 M
B   1.0 M
C   2.0 M
D   4.0 M
E   8.0 M   (L)

36 Silver chloride is precipitated when solutions of sodium chloride and silver sulphate are mixed:

$$2NaCl(aq) + Ag_2SO_4(aq)$$
$$\rightarrow 2AgCl(s) + Na_2SO_4(aq)$$

How many cm^3 of 1.0 M sodium chloride would just react with $1\,dm^3$ of 0.01 M silver sulphate?

A   1
B   2
C   10
D   20
E   100   (L)

37 What volume of 0.2 M hydrochloric acid solution would be required to neutralize $20\,cm^3$ of an alkaline solution which contains 4.0 g NaOH per dm^3?
   ($H = 1$, $O = 16$, $Na = 23$)

A   40 cm^3
B   30 cm^3
C   20 cm^3
D   10 cm^3
E   4 cm^3 (NISEC)

38 Potassium carbonate reacts with sulphuric acid according to the equation

$$K_2CO_3(aq) + H_2SO_4(aq)$$
$$\rightarrow K_2SO_4(aq) + H_2O(l) + CO_2(g)$$

What mass of carbon ($C = 12$) would have to be burnt completely in excess oxygen to yield the same mass of carbon dioxide as could be obtained by reacting $750\,cm^3$ of 0.1 M potassium carbonate solution with excess sulphuric acid?

A   0.60 g
B   0.75 g
C   0.80 g
D   0.90 g
E   1.20 g   (L)

39 When some molten lead bromide was electrolysed 0.4 mole of Br_2 was produced. How many moles of electrons had passed?

A   0.1
B   0.2
C   0.4
D   0.8
E   1.6   (NISEC)

40 The number of moles of electrons (faradays) needed to deposit 26 g of chromium ($Cr = 52$) from a solution containing $Cr^{3+}(aq)$ is

A   0.5
B   1.0
C   1.5
D   3.0
E   6.0   (L)

41 965 coulombs of electricity are passed through different cells containing $Ag^+(aq)$ ions, $Cu^{2+}(aq)$ ions and $Cr^{3+}(aq)$ ions. What mass of each metal, in grams, is deposited on the cathode of each cell?

(Ag = 108, Cu = 64, Cr = 52; the charge on 1 mole of electrons is 96 500 coulombs)

	Silver	Copper	Chromium	
A	0.36	0.32	0.52	
B	1.08	0.32	0.17	
C	1.08	0.64	0.52	
D	1.08	1.28	1.56	
E	3.24	1.28	0.52	(L)

42 During the electrolysis of an aqueous solution of a cerium salt, 70 g of cerium (Ce = 140) is deposited at the cathode by 2 moles of electrons (faradays). The formula of the cerium ion is probably

A Ce^+
B Ce^-
C Ce^{2+}
D Ce^{4+}
E Ce^{4-} (L)

43 A metallic element X (relative atomic mass 52) is deposited at the cathode when its molten chloride is electrolysed. If 0.52 g of X is deposited by the passage of 0.03 of a mole of electrons (faraday), the formula of the chloride is

A X_3Cl
B X_3Cl_2
C X_3Cl_3
D X_2Cl_3
E XCl_3 (L)

44 The heat of vaporization of ethanol (C_2H_5OH) at its boiling point is $38.7 \, kJ \, mol^{-1}$. If heat energy is supplied to ethanol at the rate of $3.87 \, kJ$ per minute, how many minutes will it take to convert 9.2 g of ethanol at its boiling point to vapour?
$(C_2H_5OH = 46)$

A 0.5
B 0.92
C 2
D 10
E 50   (L)

45 When 0.01 mole of propanol is completely burned, the heat evolved raises the temperature of 160 g of water by 30 °C.

1 kg of water requires 4.2 kJ to raise the temperature through 1 °C.

ΔH for this reaction is

$$kJ \, mol^{-1}$$

A $-\dfrac{160 \times 30 \times 4.2}{0.01 \times 1000}$

B $-\dfrac{160 \times 4.2}{30 \times 0.01 \times 1000}$

C $-\dfrac{30 \times 4.2}{160 \times 0.01 \times 1000}$

D $+\dfrac{160 \times 30 \times 4.2}{0.01 \times 1000}$

E $+\dfrac{160 \times 4.2}{30 \times 0.01 \times 1000}$ (L)

46 The heats of combustion of the simplest alcohols are shown below.

		Heat of combustion $(kJ \, mol^{-1})$
Methanol	CH_3OH	-718
Ethanol	C_2H_5OH	-1380
Propanol	C_3H_7OH	-2020

The heat of combustion of butanol, C_4H_9OH, would be about

A $-2310 \, kJ \, mol^{-1}$
B $-2440 \, kJ \, mol^{-1}$
C $-2650 \, kJ \, mol^{-1}$
D $-2860 \, kJ \, mol^{-1}$
E $-2940 \, kJ \, mol^{-1}$ (L)

47 The following quantities of sodium thiosulphate solution were added to five separate $10 \, cm^3$ portions of $2 \, M$ hydrochloric acid. The mixture was made up to a total volume of $50 \, cm^3$ in each case and the rate of reaction was determined. In which example does the reaction proceed the fastest?

A $10 \, cm^3$ of $2 \, M$ sodium thiosulphate solution

B $20 \, cm^3$ of $2 \, M$ sodium thiosulphate solution

C $10 \, cm^3$ of $3 \, M$ sodium thiosulphate solution

D $20 \, cm^3$ of $3 \, M$ sodium thiosulphate solution

E $10 \, cm^3$ of $4 \, M$ sodium thiosulphate solution   **(L)**

48 After an experiment to investigate the rate of reaction between calcium carbonate and hydrochloric acid, a graph was plotted showing the volume of gas evolved against time.

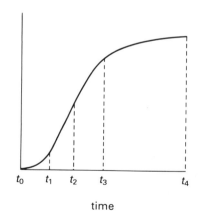

At which time was the rate of evolution of gas greatest?

A t_0

B t_1

C t_2

D t_3

E t_4 **(L)**

49 A radioactive element with a half-life of 3.5 days is used in an experiment in a research laboratory for 14 days. What fraction of the mass of the original element will be left at the end of the experiment?

A $\frac{1}{32}$

B $\frac{1}{16}$

C $\frac{1}{8}$

D $\frac{1}{4}$

E $\frac{3}{4}$ **(L)**

50 A radioactive substance which initially gave a count rate of 1200 per minute in a radiation counter gave only 150 per minute after three hours. Two pupils estimated count rates at various other times as follows:

	First pupil	Second pupil
After 1 hour	850	600
After 2 hours	500	300
After 4 hours	Zero	Zero

Which two of their estimates are correct?

A   850 and 500

B   850 and zero

C   600 and 500

D   600 and 300

E   300 and zero   **(L)**

Answers

Chapter 1

1 (a) 4 protons, 5 neutrons, 4 electrons   (b) 19 protons, 20 neutrons, 19 electrons   (c) 92 protons, 143 neutrons, 92 electrons   (d) 53 protons, 74 neutrons, 54 electrons   (e) 20 protons, 20 neutrons, 18 electrons

2 (a) $^{16}_{8}O$ (b) $^{40}_{18}Ar$ (c) $^{23}_{11}Na^{+}$ (d) $^{32}_{16}S^{2-}$

3 (a) sodium, magnesium, aluminium, silicon, phosphorus, sulphur, chlorine, argon

4 (a) 12   (b) 24   (c) 80   (d) 137   (e) 207

5 (a) 2   (b) 2   (c) 8   (d) 2   (e) 4

6 69.8

7 121.86

8 (a) 23 g   (b) 355 g   (c) 12.7 g   (d) 28 g   (e) 10 g

9 (a) 1 mol   (b) 0.5 mol   (c) 0.05 mol   (d) 10 mol   (e) 0.01 mol

10 (a) 12 g   (b) 8 g   (c) 0.64 g   (d) 19 kg

11 (a) 20 mol   (b) 0.1 mol   (c) 0.001 mol

12 (a) 32   (b) 44   (c) 128   (d) 170

13 (a) 71 g   (b) 36 g   (c) 128 g   (d) 31 g

14 (a) 0.1 mol   (b) 2 mol   (c) 0.5 mol   (d) 1 mol   (e) 0.2 mol

15 (a) 1 g   (b) 9 g

16 (a) 74.5   (b) 267   (c) 400   (d) 287   (e) 286

17 (a) 97 g   (b) 165.5 g   (c) 710 g   (d) 27.8 g   (e) 0.535 g

18 (a) 2 mol   (b) 4 mol   (c) 1 mol   (d) 2 mol   (e) 25 mol

19 (a) 127 g   (b) 430 g   (c) 31.5 g   (d) 0.06 g   (e) 700 g   (f) 64 g   (g) 12 g

20 (a) 4 mol   (b) 0.25 mol   (c) 0.25 mol   (d) 0.25 mol   (e) 1 mol   (f) 0.5 mol

21 (a) 2.4×10^{24} (b) 1.2×10^{24} (c) 9.6×10^{24} (d) 5.4×10^{24}

22 (a) 1.2×10^{24} (b) 9.6×10^{24} (c) 1.2×10^{24} (d) 5.4×10^{24}

Examination questions

1 11 protons, 22 neutrons

2 2

3 28

4 4.8 g

5 0.0083 mol

6 3 g

7 256 g

8 0.3 mol

9 284 g

10 0.48 g

11 0.168 g

12 1.6 mol

13 2 mol

14 5 mol

15 0.001 25 mol

16 0.04 mol

17 3×10^{22}
18 1.2×10^{23}
19 (a) 0.33 mol (b) 0.17 mol (c) 0.125 mol (d) 0.1 mol
20 (a) 23 11 11 11 12
             40 20 18 20 20
             37 17 18 17 20
      (c) 1 proton, 0 neutrons, 0 electrons
21 (c) 2, 8, 7 (d) 70, 72, 74
22 (b) $^{20}_{10}Ne$: 10 protons, 10 neutrons $^{22}_{10}Ne$: 10 protons, 12 neutrons
23 1.6 mol
24 (a) 5×10^{22} (b) 6×10^{23}

Chapter 2

1 Na_2O
2 (a) Mg_3N_2 (b) CH_4 (c) SiO_2 (d) $FeBr_3$
3 (a) $CaCO_3$ (b) Na_2SO_4 (c) CH_2O (d) NH_3O
4 (a) CH_2O (b) C_2H_5 (c) C_2H_6O (d) H_2SO_4
5 empirical formula CH_3O, molecular formula $C_2H_6O_2$
6 empirical formula SCl, molecular formula S_2Cl_2
7 empirical formula C_2H_5, molecular formula C_4H_{10}
8 $CaSO_4 \cdot 2H_2O$
9 1.5
10 $Na_2CO_3 \cdot 10H_2O$
11 (a) 82.35% (b) 35.00%
12 (a) 88.89% (b) 57.66% (c) 63.49%

Examination questions

1 (a) 0.16 g, 0.20 g (f) 0.16 g (g) 16 g (h) MO
2 (a) 1.12 g (b) 112 g (c) Fe_2O_3
3 Cu_2O
4 KO_2
5 PbO_2
6 (a) V_2O_3 (b) VO
7 (a) empirical formula Sb_2O_5, molecular formula Sb_2O_5
      (b) (i) empirical formula SbO_2, molecular formula Sb_2O_4
8 65
9 H_2SO_3
10 PbO_2
11 MCl_2
12 (a) C_4H_{10}
13 (a) CH_2Cl (b) $C_2H_4Cl_2$
14 (a) 28 (b) C_2H_4
15 (a) C_5H_4 (b) $C_{10}H_8$
16 (a) empirical formula NH_2, molecular formula N_2H_4
17 (a) 1:7 (b) 246
18 10
19 $Na_2CO_3 \cdot H_2O$
20 6

21   40.5%
22   12 tonnes
23   3.9%
24   (a) 21.2%   (b) 66.0%
25   (a) (i) L (ii) 0.5 L (iii) 0.33 L (b) (i) 60 (ii) 0.01 mol (iii) 6×10^{21}

Chapter 3

1   2.4 g
2   14 g
3   (a) 8.4 tonnes   (b) 11.2 tonnes
4   1.2 tonnes
5   0.45 g
6   68 g
7   414 g
8   2.0 g
9   3.2 g
10   7.3 g
11   4.14 g
12   0.81 g
13   5.3 g
14   37.6 g

Examination questions

1   (a) (i) 40 (ii) 120   (b) 3 g
2   22 tonnes of carbon dioxide, 28 tonnes of calcium oxide
3   56.5 g
4   (b) 34.8 kg
5   (a) 56 tonnes
6   (a) (ii) 16.1 g   (b) 5.825 g of barium sulphate
7   8.52 g
8   (b) 314 million tonnes
9   490 tonnes
10   (a) 500 tonnes
11   120 tonnes
12   (b) 460 kg   (c) 87%
13   (b) 44 g   (c) 75%
14   (a) 0.64 kg   (b) 0.064 kg

Chapter 4

1   (a) 2000 cm^3 (b) 50 cm^3 (c) 500 cm^3 (d) 4550 cm^3 (e) 568 cm^3
2   (a) 5 dm^3 (b) 2.5 dm^3 (c) 0.5 dm^3 (d) 1000 dm^3
3   (a) 718 K   (b) 1813 K   (c) 90 K   (d) 266 K
4   (a) 24 dm^3 (b) 50 cm^3 (c) 67 cm^3 (d) 48 cm^3 (e) 38 cm^3
5   (a) 50 cm^3 (b) 62.5 cm^3 (c) 731 cm^3 (d) 112 cm^3
6   (a) 100 cm^3 (b) 460 dm^3 (c) 514 cm^3

7   (a) 1 mol   (b) 0.002 mol   (c) 10 mol   (d) 0.125 mol   (e) 3 mol
8   (a) 1 mol   (b) 0.004 mol   (c) 10 mol   (d) 0.005 mol
9   (a) (i) $44.8 \, dm^3$ (ii) $48 \, dm^3$ (b) (i) $224 \, dm^3$ (ii) $240 \, dm^3$ (c) (i) $224 \, cm^3$
      (ii) $240 \, cm^3$ (d) (i) $0.0448 \, cm^3$ (ii) $0.048 \, cm^3$ (e) (i) $2800 \, cm^3$ (ii) $3000 \, cm^3$
10   (a) (i) $22.4 \, dm^3$ (ii) $24.0 \, dm^3$ (b) (i) $2240 \, cm^3$ (ii) $2400 \, cm^3$
      (c) (i) $2.24 \, cm^3$ (ii) $2.4 \, cm^3$ (d) (i) $5.6 \, cm^3$ (ii) $6.0 \, cm^3$ (e) (i) $0.0224 \, cm^3$
      (ii) $0.024 \, cm^3$
11   $52 \, g$
12   $38 \, g$
13   (b) 1   (c) 2   (d) $44 \, g$ (e) N_2O
14   $28 \, g$, CO
15   (a) CH_3 (b) 30   (c) C_2H_6
16   $224 \, cm^3$
17   (a) $3000 \, cm^3$ (b) 0.125 mol   (c) 0.125 mol   (d) $12.5 \, g$
18   $240 \, cm^3$
19   $84.3 \, cm^3$
20   (a) $100 \, cm^3$ (b) $2 \, cm^3$ (c) $103 \, dm^3$
21   $175 \, cm^3$ of oxygen, $100 \, cm^3$ of carbon dioxide
22   $25 \, cm^3$ of oxygen, $50 \, cm^3$ of nitrogen dioxide
23   $20 \, cm^3$ of hydrogen bromide
24   $24 \, cm^3$
25   C_2H_6
26   C_6H_{12}

Examination questions

1   (a) A: 17 protons, 17 electrons, 18 neutrons
         B: 17 protons, 17 electrons, 20 neutrons
      (b) 35.5   (c) $120 \, cm^3$ (d) 313°C   (e) 2
2   (c) $0.134 \, g$
3   (a) neon 37 s, oxygen 47 s, sulphur dioxide 60 s   (b) C_4H_{10} (c) CO, C_2H_4
4   (b) 0.0025 mol   (c) 0.0025 mol   (d) $0.06 \, g$ (e) 60%
6   89.6%
7   (a) $5600 \, cm^3$ (b) $5600 \, cm^3$
8   (a) $44\,800 \, dm^3$ (b) $88 \, kg$
9   (b) (i) $13.2 \, g$ (ii) $2.8 \, dm^3$
10   $2800 \, dm^3$
11   $28 \, dm^3$ of oxygen, $22.4 \, dm^3$ of nitrogen monoxide
12   $4.25 \, g$
13   $67.2 \, dm^3$, $11.2 \, dm^3$
14   $4.48 \, dm^3$
15   $112 \, cm^3$ of ethene, $88 \, cm^3$ of methane
16   (a) $4.48 \, dm^3$ (b) $1.12 \, dm^3$
17   $654 \, cm^3$
18   (a) $715 \, dm^3$ (b) $715 \, dm^3$ (c) $1287 \, dm^3$
19   $2 \, dm^3$
20   $89.6 \, dm^3$
21   $90 \, g$
22   (a) 28   (b) C_2H_4 (c) $200 \, cm^3$
23   (a) CH_2Br (b) 188   (c) molecular formula $C_2H_4Br_2$
24   (a) C_3H_7 (b) (i) $86 \, g/mol$ (ii) C_6H_{14}
25   (b) $22.4 \, dm^3$

26   molecular formula Kr
27   total volume $40\,cm^3$: $20\,cm^3$ of oxygen and $20\,cm^3$ of nitrogen
28   (b) (i) $200\,cm^3$ (ii) total volume $180\,cm^3$:$160\,cm^3$ of nitrogen and $20\,cm^3$ of carbon dioxide
29   (b) (i) $50\,cm^3$ (ii) $55\,cm^3$
30   (c) 1.5×10^{22} (d) $2500\,cm^3$
31   (b) (i) $0.001\,25\,mol$ (ii) $0.001\,25\,mol$ (iii) $0.015\,g$ (iv) $0.001\,g$ (v) $0.001\,mol$
         (c) (i) C_7H_5 (ii) $C_{14}H_{10}$
32   (b) $20\,cm^3$ (c) $40\,\%$
33 $40\,cm^3$ of carbon dioxide, $30\,cm^3$ of oxygen
34   4
35   (a) C_3H_8O (b) two —OH groups

Chapter 5

1 $36.60\,g/100\,g$
2 $203.87\,g/100\,g$
3   (a) $20\,°C$ $37.17\,g/100\,g$
              $40\,°C$ $45.80\,g/100\,g$
              $60\,°C$ $55.18\,g/100\,g$
              $80\,°C$ $65.60\,g/100\,g$
4   (a) (i) $16.67\,g/100\,g$ (ii) $12.50\,g/100\,g$ (iii) $10.00\,g/100\,g$
5   (a) $88\,°C$ $200.0\,g/100\,g$
              $56\,°C$ $100.0\,g/100\,g$
              $41\,°C$ $66.7\,g/100\,g$
              $32\,°C$ $50.0\,g/100\,g$
              $25\,°C$ $40.0\,g/100\,g$
              $21\,°C$ $33.3\,g/100\,g$
6   (b) $37.2\,g/100\,g$ (c) no   (d) $40\,°C$ (e) $200\,cm^3$ (f) $3.4\,g$ (g) $11.6\,g$ (h) $140\,g$
         (i) $2.9\,g$
7   (a) $72.0\,g$ (b) $7.6\,g$
8   (b) $23.5\,g$ (c) $10.35\,g$ (d) $13.15\,g$ (e) $18.3\,g$ (f) $2\,°C$
9   (a) (i) $0.7\,g$ (ii) $2.7\,g$ (b) (i) $3.4\,g$ (ii) $5.1\,g$ (c) yes   (d) (i) $1.3\,g$ (ii) $0\,g$
10 $33.3\,\%$
11   (a) $1.0\,mol/dm^3$ (b) $2.0\,mol/dm^3$ (c) $0.1\,mol/dm^3$ (d) $0.5\,mol/dm^3$
12   (a) $0.0125\,mol$ (b) $0.2\,mol$ (c) $0.000\,05\,mol$ (d) $0.025\,mol$
13   (a) $36.5\,g$ (b) $0.04\,g$ (c) $196\,g$ (d) $0.004\,25\,g$
14   (a) (i) $4.0\,g/dm^3$ (ii) $0.1\,mol/dm^3$ (b) (i) $17.0\,g/dm^3$ (ii) $0.1\,mol/dm^3$
         (c) (i) $207.5\,g/dm^3$ (ii) $1.25\,mol/dm^3$ (d) (i) $1.58\,g/dm^3$ (ii) $0.01\,mol/dm^3$
15 $0.29\,g$
16 $2.24\,dm^3$
17 $2.0\,g$ of copper(II) oxide, $4.0\,g$ of copper(II) sulphate
18 $4.62\,g$
19   (a) $5.0\,g$ (b) $1.12\,dm^3$
20 $10\,cm^3$
21 $Pb(NO_3)_2(aq) + 2NaI(aq)$
22   (b) $2.0\,mol/dm^3$ (c) $20\,mol/dm^3$
23   (b) $0.002\,mol$ (c) $0.004\,mol$ (d) $0.08\,mol$
24 $0.108\,mol/dm^3$

25 (a) barium chloride $0.5\,\text{mol/dm}^3$, potassium chromate(VI) $0.5\,\text{mol/dm}^3$
    (d) $BaCl_2(aq) + K_2CrO_4(aq)$
26 (a) $2KCl(aq) + Pb(NO_3)_2(aq)$ (c) 2.78 g
27 (a) $2Na_2S_2O_3(aq) + I_2(aq)$
28 (a) $H_xY(aq) + 3NaOH(aq)$ (b) 3
29 (a) $2HNO_3(aq) + K_2CO_3(aq)$ (b) 1 mol
30 (a) $0.17\,\text{mol/dm}^3$ (b) $10.5\,\text{g/dm}^3$
31 $0.22\,\text{mol/dm}^3$
32 (a) $0.05\,\text{mol/dm}^3$ (b) 126 g, $x = 2$

Examination questions

1   $32.4\,\text{g}/100\,\text{g}$
3   (b) 18 g
4   (c) 51 °C   (d) 39 °C   (e) (i) 40 g (ii) 30 g (iii) 10 g
5   (b) (i) $49.5\,\text{g}/100\,\text{g}$ (iii) 51 °C   (v) 73.5 %
6   39 g
7   (a) (iii) 33.3 % oxygen, 66.7 % nitrogen
8   (a) 720 g, 19.7 mol   (b) 19.7 mol/litre
9   0.042 mol
10 (a) 6300 mol   (b) 2.52 tonnes
11 (c) $0.174\,\text{g/dm}^3$ (d) 87 %
12 $5.5\,\text{mol/dm}^3$
13 46.6 g
14 $125\,\text{cm}^3$
15 (a) 4.2 g   (b) $1.12\,\text{dm}^3$
16 2.1 g, $0.56\,\text{dm}^3$
17 (b) $\frac{1}{20}$ (c) $\frac{1}{10}$ (d) 2   (e) MCl_2, $MCl_2(aq) + 2AgNO_3(aq)$
    $\rightarrow 2AgCl(s) + M(NO_3)_2(aq)$ (f) 40
18 (b) 0.004 mol   (c) 0.004 mol   (d) 74 g/mol   (e) 2
19 (a) $30.0\,\text{cm}^3$ (b) $0.2\,\text{mol/dm}^3$ (c) $0.24\,\text{mol/dm}^3$ (d) 46
20 (b) (i) $0.18\,\text{mol/dm}^3$ (ii) 7.2 g/litre
21 (a) 0.01 mol   (b) 0.02 mol   (c) 0.5 mol   (d) $18.25\,\text{g/dm}^3$ (e) $20\,\text{cm}^3$
22 (a) $0.2\,\text{mol/dm}^3$ (b) $34.2\,\text{g/dm}^3$
23 (a) 0.05 mol   (b) 0.025 mol   (c) 184 g/mol   (d) 24
24 (a) 0.0089 mol   (b) 0.0089 mol   (c) 0.0089 mol   (d) 0.205 g, 10.2 %

Chapter 6

2   (a) 1 mol   (b) 2 mol   (c) 2 mol   (d) 3 mol
3   (a) 0.2 mol   (b) 0.5 mol   (c) 1.0 mol   (d) 0.3 mol
4   (a) 0.2 mol   (b) 0.1 mol   (c) 0.1 mol   (d) 0.067 mol
5   (a) 4.6 g   (b) 6.4 g   (c) 5.9 g   (d) 3.73 g
6   (a) 1 mol   (b) 0.2 mol   (c) 0.2 mol   (d) 4 mol
7   (a) $24\,\text{dm}^3$ (b) $1.2\,\text{dm}^3$ (c) $12\,\text{dm}^3$ (d) $2.4\,\text{dm}^3$
8   (a) 1 mol   (b) 0.2 mol   (c) 0.01 mol   (d) 0.02 mol
9   (a) 0.1 mol   (b) 0.25 mol   (c) 0.05 mol
10 (a) $4.8\,\text{dm}^3$ (b) $1.2\,\text{dm}^3$ (c) $0.24\,\text{dm}^3$
11 (a) 25.4 g   (b) 0.254 g   (c) 1.27 g
12 (a) 0.25 mol   (b) 0.02 mol   (c) 0.8 mol   (d) 0.2 mol

13 (a) $24\,cm^3$ (b) $600\,cm^3$ (c) $1200\,cm^3$
14 (b) $0.0005\,mol$ (c) $0.032\,g$
15 (a) $120\,C$ (b) $600\,C$ (c) $3600\,C$
16 (a) $500\,s$ (b) $1250\,s$ (c) $2000\,s$
17 (a) $19\,300\,C$ (b) $386\,C$ (c) $4825\,C$
18 (a) $0.02\,mol$ (b) $0.005\,mol$ (c) $0.001\,mol$ (d) $0.3\,mol$
19 (a) $0.002\,mol$ (b) $0.005\,mol$ (c) $0.006\,mol$ (d) $0.010\,mol$
20 (a) $0.005\,mol$ (b) $0.01\,mol$ (c) $0.0033\,mol$ (d) $0.01\,mol$
21 (a) $2.07\,g$ (b) $0.59\,g$ (c) $2.16\,g$ (d) $0.37\,g$
22 (a) $965\,C$ (b) $0.01\,mol$ (d) $0.005\,mol$ (e) $1.035\,g$
23 $0.118\,g$
24 (a) $0.192\,g$ (b) $0.192\,g$
25 (a) $4.32 \times 10^9\,C$ (b) $45\,000\,mol$ (c) $405\,000\,g$ (d) $81\,tonnes$
26 (a) $965\,C$ (b) $0.01\,mol$ (d) $0.005\,mol$ (e) $120\,cm^3$
27 (a) $120\,cm^3$ (b) $60\,cm^3$
28 (a) $0.02\,mol$ (b) $2\,mol$ (c) $2+$
29 (a) $386\,C$ (b) $0.004\,mol$ (c) $0.002\,mol$ (d) $2\,mol$ (e) $2+$
30 $3+$
31 (a) $9650\,C$ (b) $38\,600\,C$ (c) $2895\,C$ (d) $19\,300\,C$
32 (a) $96.5\,s$ (b) $193\,s$ (c) $289.5\,s$
33 (a) $0.002\,mol$ (b) $0.004\,mol$ (c) $386\,C$ (d) $772\,s$
34 (a) $0.004\,mol$ (c) $0.008\,mol$ (d) $772\,C$ (e) $0.8\,A$
35 $0.016\,A$
36 $19\,300\,s$
37 $1544\,s$
38 (a) $0.001\,mol$ (c) $0.002\,mol$ (e) $0.002\,mol$ (f) $0.216\,g$
39 (a) $0.004\,mol$ · (c) $0.008\,mol$ (e) $0.004\,mol$ (f) $96\,cm^3$
40 (a) $0.005\,mol$ (b) $0.010\,mol$ (c) $965\,C$ (d) $0.8\,A$ (f) $0.003\,mol$ (g) $0.17\,g$
(h) 30 minutes
41 (a) $0.002\,mol$ (b) $0.002\,mol$ (c) $2\,mol$ (d) $2+$
42 $60\,cm^3$

Examination questions

1 (d) (i) $48\,C$ (ii) $0.051\,g$ (iii) $5.97\,cm^3$
2 (a) $35.5\,g$ of chlorine   (b) $23\,g$ of sodium
3 (d) (i) $300\,000\,C$ (e) $1.8\,dm^3$
4 (c) (i) $193\,C$ (ii) $2\,mol$ (e) 24 minutes
6 (c) $3860\,s$ (d) $0.032\,g$ (e) $100\,g$, M^{2+}
7 (b) $960\,C$ (c) $0.005\,mol$ (d) $2\,mol$
8 (e) $0.094\,mol/dm^3$
9 (f) (i) $2\,mol$ (ii) MCl_2
10 (a) (i) $3\,mol$, 3
11 (a) cathode: $11.2\,dm^3$ of hydrogen
        anode: $5.6\,dm^3$ of oxygen
12 (e) (ii) $0.002\,mol$ (iii) $0.001\,mol$
13 (d) $8.5\,g$ (e) (ii) $0.01\,mol$ (iii) $960\,C$ (iv) $1\,mol$

Chapter 7

1   (a) 10 kJ   (b) 0.3 kJ   (c) 1.2 kJ
2   (a) 15 000 J   (b) 250 J   (c) 2500 J
3   (a) 4000 J   (b) 6000 J   (c) 360 kJ
4   20 W   (b) 40 W   (c) 2.5 kW
5   (a) 1050 J   (b) 600 J
6   5 °C
7   28 °C
8   336 kJ
9   (a) 50 400 J   (b) 280 J/s
10   5 °C
11   420 s
12   2.5 J/(g °C)
13   8.4 °C
14   28.6 kJ
15   (a) 46 g   (b) 0.1 mol   (c) 39 kJ
16   (a) 160 g   (b) 0.025 mol   (c) 30 kJ
17   (a) 84 g   (b) 0.025 mol   (c) 35 kJ
18   (a) 1600 J   (b) 58 g   (c) 0.05 mol   (d) 32 kJ
19   (a) 100 J   (b) 5 s
20   (a) water   (b) 322 J   (c) 32.2 kJ
21   20.5 kJ
22   460 kJ
23   5 kJ
24   (a) 6.4 kJ   (b) 320 kJ
25   480 kJ
26   (a) 8400 J   (b) 120 J/s   (c) 12 000 J   (d) 100 s
27   3920 kJ
28   (a) 72 g   (b) 0.05 mol   (c) 2440 kJ
29   4850 kJ
30   (a) 17.85 kJ   (b) 0.02 mol   (c) 892.5 kJ
31   (a) 8.4 kJ   (b) 0.005 mol   (c) 1680 kJ
32   (a) 24.5 kJ   (b) 0.02 mol   (c) 1225 kJ
33   (a) 35.7 kJ   (b) 2642 kJ
34   (a) hydrogen   (b) methane
35   (b) (i) 3500–3600 kJ (ii) 4800–4900 kJ
36   (a) 20.2 kJ   (b) 24 °C
37   (a) 32.75 kJ   (b) 32.75×10^6 kJ
38   (a) 168 kJ   (b) 0.2 mol   (c) 4800 cm^3
39   (b) 0.01 mol   (c) 5040 J   (d) 504 kJ
40   (b) 0.02 mol   (e) 0.2 mol   (d) zinc   (e) 2520 J   (f) 126 kJ
41   157.5 kJ
42   (b) 0.05 mol   (c) 1050 J   (d) 21 kJ
43   (a) 0.2 mol   (b) 3360 J   (c) 16.8 kJ   (d) 4 °C
44   (c) (i) 10.5 kJ absorbed (ii) 63.0 kJ released   (d) 50 °C
45   (a) 0.5 mol   (b) 0.5 mol/dm^3 (c) 33.6 kJ   (d) 67.2 kJ   (e) 32 °C
46   (b) 0.01 mol   (c) 1890 J   (d) 189 kJ
47   (a) 0.005 mol   (b) 735 J   (c) 147 kJ
48   (a) 0.005 mol   (b) 840 J   (c) 336 kJ
49   (a) 0.005 mol   (b) 0.005 mol   (c) 210 J   (d) 42 kJ   (e) 1 °C

50 (b) 18.5 °C (c) − 1 °C (e) 210 J (f) 0.025 mol (g) 0.025 mol (h) 8.4 kJ
51 (b) 0.025 mol (c) 1365 J (d) 54.6 kJ
52 (b) 109.2 kJ
53 (d) 54.6 kJ
54 (b) − 6 kJ/mol
55 (b) + 41 kJ/mol
56 (a) 0.1 mol (c) + 39 kJ/mol
57 (a) 0.01 mol (c) − 4195 kJ/mol
58 (b) − 460 kJ/mol
59 (b) 21 kJ (c) + 21 kJ/mol
60 (b) − 140 kJ/mol
61 (e) 184 kJ/mol released
62 − 103 kJ/mol
63 (b) − 93 kJ/mol

Examination questions

1 (c) (i) 200 kJ (ii) 30.8 kJ/mol
2 (c) 10 minutes (d) 0.5 mol
3 (d) 29.73 kJ (e) (i) 150 kJ (ii) 377.4 g
4 (a) (iii) 20 litres (iv) 100 litres (vi) 11 kg (b) (i) 600 cm^3 (ii) 313 °C
5 (b) (ii) ethanol − 1300 to − 1400 kJ/mol, hexanol − 3900 to − 400 kJ/mol
6 (d) 0.02 mol (e) 21 kJ (f) − 1050 kJ/mol
7 (a) 4 minutes (c) 80 °C (d) 32 000 J (e) 32 kJ (f) 8 kJ
8 (c) (i) − 1365 kJ/mol (d) (i) 0.007 mol (ii) 0.322 g
9 (a) 49 g (b) − 71.4 kJ/mol
10 (d) (i) 1050 J (ii) − 210 kJ/mol
11 (c) 0.02 mol of bromine *atoms*, $HgBr_2$ (h) − 172 kJ/mol
12 (e) − 54.6 kJ/mol

Chapter 8

1 (a) (i) 0.0048 g/s (ii) 0.0002 mol/s (b) 0.0004 mol/s
2 (a) (i) 7.2 cm^3/s (ii) 0.0003 mol/s (b) 0.0003 mol/s (c) 0.03 g
3 (a) (i) 0.002 g/s (ii) 0.000 05 mol/s (b) (i) 0.000 05 mol/s (ii) 1.2 cm^3/s
4 (a) (i) 0.26 cm^3/s (ii) 0.12 cm^3/s (iii) 0.08 cm^3/s (iv) 0.02 cm^3/s (v) 0.01 cm^3/s
(vi) 0.01 cm^3/s (d) 400 s
5 (b) (i) 0.015 g/s (ii) 0.006 g/s (d) (i) 480 s (ii) 600 s
6 (b) 0.0073 g/s (d) about twice as fast (f) (i) 0.08 mol (ii) 0.04 mol (h) (i) 1.60 g
(ii) 0.40 g
7 (c) 24 s (d) 0.017/s, 0.033/s, 0.05/s, 0.067/s, 0.083/s
8 (a) 60 °C (c) (i) halved (ii) halved (d) 0.0038/s, 0.0077/s, 0.015/s, 0.032/s,
0.062/s (g) × 2
9 (a) (i) faster (ii) the same (b) (i) slower (ii) the same (c) (i) faster (ii) × 2
(d) (i) slower (ii) the same
11 (a) 0.005 mol (b) 120 cm^3 (d) 90 s
12 (c) 25 minutes
13 (c) 15 hours
14 (a) 140 counts/minute (b) 70 counts/minute (c) 35 counts/minute
15 (a) 20 counts/s (b) 60 hours from the start

16 17 190 years
17 0.0005 g
18 (b) 0.1 g   (c) 71 years

Examination questions

1   (a) (ii) 60 cm³, 13 s
2   (b) about 450 cm³   (d) about 0.8 mol/dm³
4   (b) (i) 50 cm³ (ii) 30 s   (e) (ii) 0.01 g
5   (a) initially   (b) 120 s   (c) 48 cm³   (d) 29 s   (e) 0.002 mol   (f) 4 cm³
8   (a) about 74 cm³   (d) 74 cm³
9   (c) 66 cm³/minute   (f) (i) 12 040 cm³
10 (c) (i) 210 s
11 (b) P 0.115 g, Q 0.20 g   (c) (i) 5 s (ii) 20 s
12 (c) 61 cm³   (d) (i) 0.0026 mol (ii) 0.0026 mol   (e) 63 cm³   (f) 0.0052 mol
       (g) 10.4 cm³
13 (d) (ii) 180 s (iii) 70 cm³ (iv) 0.5 g (v) 7 volume
14 (b) 1.67 mol   (e) 200 cm³
15 (b) (i) 60 cm³ (ii) 35 cm³ (iii) 0 cm³
16 (c) 65 cm³   (g) (i) 13 cm³ (ii) about 62 s
17 (a) (i) about 38 s (ii) about 14 s   (c) 0.064 g
18 (d) about 74 s   (e) 60 s
19 (b) about 135 s
20 (d) about 60 minutes   (e) 0.001 25 g   (f) (i) $^{234}_{90}X$, $^{234}_{91}Y$, $^{234}_{92}Z$
21 (a) (i) $^{228}_{88}B$, $^{216}_{84}G$ (c) (i) 375 counts/minute (iii) about 1070 counts/minute
22 (a) about 350 g   (b) about 185 g
23 (c) 2.6×10^9 years

Multiple choice test

1 D	2 A	3 C	4 A	5 E	6 B	7 A	8 D
9 D	10 D	11 E	12 C	13 E	14 C	15 D	16 E
17 E	18 D	19 A	20 C	21 E	22 D	23 D	24 D
25 C	26 E	27 B	28 B	29 C	30 C	31 C	32 D
33 D	34 B	35 C	36 D	37 D	38 D	39 D	40 C
41 B	42 D	43 E	44 C	45 A	46 C	47 D	48 C
49 B	50 D